Information Science and Knowledge Management, Vol. 15

Peter Bruza · Marc Weeber
Editors

Literature-based Discovery

With 25 Figures and 29 Tables

Editors
Prof. Dr. Peter Bruza
Queensland University of Technology
Brisbane 4001
GPO Box 2434
Australia
p.bruza@qut.edu.au

Dr. Marc Weeber
733 Fallsgrove Dr.
Rockville MD 20850
USA
marc@weeber.net

ISBN: 978-3-642-08819-3 e-ISBN: 978-3-540-68690-3

Softcover reprint of the hardcover 1st edition 2008

Cover design: KünkelLopka, Heidelberg

Printed on acid-free paper

5 4 3 2 1 0

springer.com

Preface

When Don Swanson hypothesized a connection between Raynaud's phenomenon and dietary fish oil, the field of literature-based discovery (LBD) was born. During the subsequent two decades a steady stream of researchers have published articles about LBD and the field has made steady progress in laying foundations and creating an identity. It is curiously significant that LBD is not "owned" by any particular discipline, for example, knowledge discovery or text mining. Rather, LBD researchers originate from a range of fields including information science, information retrieval, logic, and the biomedical sciences. This reflects the fact LBD is an inherently multi-disciplinary enterprise where collaborations between the information and biomedical sciences are readily encountered. This multi-disciplinary aspect of LBD has made it harder for the field to plant a flag, so to speak. The present volume can be seen as an attempt to redress this. It presents chapters providing a broad brush stroke of LBD by leading researchers providing an overview of the state of the art, the models and theories used, experimental studies, lessons learnt, application areas, and future challenges. In short, it attempts to convey a learned impression of where and how LBD is being deployed. Don Swanson has kindly agreed to provide the introductory chapter. It is the hope and intention that this volume will plant a flag in the ground and inspire new researchers to the LBD challenge.

Peter Bruza
Marc Weeber

July 2007

Contents

Preface v

Contributors ix

Part I General Outlook and Possibilities

Literature-Based Discovery? The Very Idea 3
D.R. Swanson

The Place of Literature-Based Discovery in Contemporary Scientific Practice 13
Neil R. Smalheiser and Vetle I. Torvik

The Tip of the Iceberg: The Quest for Innovation at the Base of the Pyramid 23
M.D. Gordon and N.F. Awad

The 'Open Discovery' Challenge 39
Jonathan D. Wren

Where is the Discovery in Literature-Based Discovery? 57
R.N. Kostoff

Part II Methodology and Applications

Analyzing LBD Methods using a General Framework 75
A.K. Sehgal, X.Y. Qiu, and P. Srinivasan

Evaluation of Literature-Based Discovery Systems 101
M. Yetisgen-Yildiz and W. Pratt

Factor Analytic Approach to Transitive Text Mining using Medline Descriptors 115
J. Stegmann and G. Grohmann

Literature-Based Knowledge Discovery using Natural Language Processing 133
D. Hristovski, C. Friedman, T.C. Rindflesch, and B. Peterlin

Information Retrieval in Literature-Based Discovery 153
W. Hersh

Biomedical Application of Knowledge Discovery 173
A. Koike

Index 193

Contributors

Neveen Farag Awad
School of Business, Wayne State University, USA

Carol Friedman
Department of Biomedical Informatics, Columbia University, 622 West 168 St, New York, NY 10032, USA

Michael D. Gordon
Ross School of Business, University of Michigan, USA
mdgordon@umich.edu

Guenter Grohmann
Charité-Institute of Medical Informatics, 12203 Berlin, Germany
guenter.grohmann@charite.de

William Hersh
Department of Medical Informatics and Clinical Epidemiology,
School of Medicine, Oregon Health and Science University,
3181 SW Sam Jackson Park Rd., Portland, OR 97239, USA
hersh@ohsu.edu

Dimitar Hristovski
Institute of Biomedical Informatics, Medical Faculty, University of Ljubljana, Vrazov trg 2/2, 1104 Ljubljana, Slovenia
dimitar.hristovski@mf.uni-lj.si

Asako Koike
Central Research Laboratory, Hitachi Ltd. 1-280 Higashi-Koigakubo Kokubunji City, Tokyo 185-8601, Japan

and

Department of Computational Biology, Graduate School of Frontier Science The University of Tokyo, Kiban-3A1(CB01) 5-1-5, Kashiwanoha Kashiwa, Chiba 277-8561, Japan
akoike@hgc.jp

Ronald N. Kostoff
307 Yoakum Parkway, Alexandria, VA 22304, USA
kostoff@mitre.org

Borut Peterlin
Division of medical genetics, UMC, Slajmerjeva 3, Ljubljana, Slovenia

Wanda Pratt
The Information School, University of Washington, Seattle, USA

and

Biomedical and Health Informatics, School of Medicine, University of Washington, Seattle, USA
wpratt@u.washington.edu

Xin Ying Qiu
Department of Management Sciences, The University of Iowa, Iowa City, IA 52242, USA

Thomas C. Rindflesch
National Library of Medicine, Bethesda, Maryland, USA

Aditya Kumar Sehgal
Department of Computer Science, The University of Iowa, Iowa City, IA 52242, USA

Neil R. Smalheiser
UIC Psychiatric Institute MC912, 1601 W. Taylor Street, Chicago, IL 60612, USA
neils@uic.edu

Padmini Srinivasan
Department of Computer Science, The University of Iowa, Iowa City, IA 52242, USA

and

Department of Management Sciences, The University of Iowa, Iowa City, IA 52242, USA

and

School of Library and Information Science, The University of Iowa, Iowa City, IA 52242, USA
psriniva@iowa.uiowa.edu

Johannes Stegmann
Buergipfad 24, 12209 Berlin, Germany
johannes.stegmann@onlinehome.de

Don. R. Swanson
University of Chicago, 1010 E. 59th Street, Chicago, IL 60637, USA
swanson@uchicago.edu

Veltle I. Torvik
UIC Psychiatric Institute MC912, 1601 W. Taylor Street, Chicago, IL 60612, USA

Jonathan D. Wren
Department of Arthritis and Immunology, Oklahoma Medical Research Foundation, 825 N.E. 13th Street, Room W313, Oklahoma City, Oklahoma 73104-5005, USA
jonathan.wren@ou.edu

Meliha Yetisgen-Yildiz
The Information School, University of Washington, Seattle, USA
melihay@u.washington.edu

Part I
General Outlook and Possibilities

Literature-Based Discovery? The Very Idea

D.R. Swanson

Abstract How is it possible to extract new knowledge from something already published? The possibility arises, for example, when two articles considered together suggest information of scientific interest not apparent from either article alone. In that sense, the two articles are complementary, a relationship based on the scientific problems, findings, and arguments presented. Whether the information found is also new and can lead to a plausible, testable hypothesis requires further searching and analysis of the literature from which it emerged.

The purpose of this introduction is to outline goals, concepts, problems, and literature structures that offer one approach to understanding the potential and limitations of literature-based discovery (LBD) independently of specific computer techniques that may be used to assist or implement it. The seeds of most of the basic concepts of LBD can be seen within the following classic exemplar of complementarity from a century ago that was of extraordinary importance to science.

1 Complementarity of Hybridity and Cytology: The Birth of Cytogenetics

A classic work by Walter Sutton in 1903 represents a landmark in genetics known as the Boveri–Sutton hypothesis [5]. The famous 1866 paper by Mendel on pea hybridization, that resurfaced in 1900, was interpreted by Sutton in the light of chromosome behavior observed in cell division and fertilization. An introduction to Sutton's article, written by Peters in 1959, bears a remarkable resemblance to literature-based discovery:

> "... When an author takes a series of apparently unrelated facts and ideas from two areas of investigation, combines them so that they make new sense, and develops a new hypothesis

D.R. Swanson
University of Chicago, 1010 E. 59th Street, Chicago, IL 60637, USA
swanson@uchicago.edu

P. Bruza and M. Weeber (eds.), *Literature-based Discovery,*
Springer Series in Information Science and Knowledge Management 15.

> from the combination, he not only aids in the advance of both fields but also is quite likely to open up a new one... In Sutton's paper you will see this development of relationships between the fields of cytology and heredity, which, at the time Sutton wrote, were considered to be fairly divergent from one another, in that no research techniques were shared... Sutton's paper can be considered the beginning of cytogenetics... This paper is a good model to follow in the preparation of a study involving synthesis and correlation." [5, p. 27]

To get some idea of the nature of the complementarity in Sutton's synthesis, a few salient features of Mendel's paper on hybridity and of the separate work in cytology may be helpful [8].

Mendel experimentally bred strains of peas with distinctive visible traits and found that hybrids from parents that breed true and differ consistently in one trait all look like the parent with the "dominant" trait. If those hybrids are then inbred, their first generation descendants show a 3:1 ratio of dominant to "recessive", where the latter is a reversion to the non-dominant grandparent. Continuing through one more generation, Mendel found that the recessives do not further vary, but only 2/3 of those bred from the hybrids that possess the dominant character show again a 3:1 ratio, thus indicating that the original 3:1 ratio could be reinterpreted as 1:2:1. Mendel continued his experiments through about six generations, and concluded that his results could be explained by assuming that the dominant and recessive traits split up within the new seeds and pollen, and then recombine at random during fertilization. He went on to show that a second pair of traits behaved in exactly the same way and as though entirely independent of the first pair.

Mendel's experiments entailed more than 10,000 plantings of peas. A cartoon shows a group of monks sitting at a dining table, and one monk walking in with a huge serving bowl. The caption reads: "Brother Mendel, we are getting a little tired of peas".

Turning now to the separate field of cytology (as of 1900) its basic data were derived not from plant breeding experiments but rather from examining the cell nucleus using a microscope. In the process of germ-cell division, paired chromosomes, one from each parent, separate to form gametes. Observations of the detailed behavior and orientation of chromosomes prior to dividing led Sutton to suggest that the gametes formed are just as likely to receive any given chromosome from one parent as the other. He saw that if he associated a pair of parental traits with a pair of parental chromosomes, then he could account for Mendel's observation that traits appeared to split up within the germ cells and recombine at random during subsequent fertilization. It also appeared that a second pair of traits and chromosomes behaved independently of the first pair, and so could account for Mendel's laws of segregation and distribution. Thus the problem posed by Mendel, of how a pair of traits can behave as though they were randomly distributed to progeny, is solved by Sutton's interpretation of chromosome behavior during meiosis and fertilization, which provides a causal mechanism sufficient to explain Mendel's results.

Sutton's paper holds at least two important lessons for literature-based discovery.

First, the two fields of experimental hybridity and cytology of the cell nucleus were good prospects for the analysis of complementarity even prior to 1900 because they were addressed to a common problem, in this case the transmission of hereditary traits.

Second, a more detailed study of cytology, focused on cell division and fertilization during the two decades before 1900, suggests that Sutton's synthesis was far more than a mechanistic process of putting two things together. It involved both inventiveness and substantial knowledge, both implicit and explicit. Once Sutton, who was a cytologist, had become aware of the Mendel paper, even a supercomputer of today would have been of little use in helping him create or interpret connections between cytology and hybridity. Recognizing complementarity is quintessentially a human function that requires creativity, inventiveness, scientific knowledge, and background knowledge – the latter including commonplace knowledge such as is needed for, among other things, understanding natural language in scientific text, or any other text, or to understand the point of a metaphor, a joke, or a cartoon, all of which depend on usage, context, and situation [2].

I know of no reason to believe that Sutton's achievement, notwithstanding its extraordinary importance, is unique in its dependence on human mental abilities in order to recognize complementary relationships. The achievement presents a challenge to people who understandably want the computer to do most of the work. Indeed, it seems likely that the creativity required is not unrelated to the importance of Sutton's work. If we design LBD procedures to find important connections by stimulating human creativity, the less important will follow by default, but not vice versa.

The goal of LBD in my opinion should be to support and enhance human ability by focusing on the key problems of finding promising pairs of scientific articles that can serve as a stimulus, and on identifying associated literature structures (see below). It is, in any event, plausible to assume that two articles randomly selected from a vast literature would have almost no chance of being complementary, so we need a search process that combines human knowledge and judgment with computer speed and data capacities. One key problem here is to determine what kind of clues are helpful in pointing to or defining "promising" pairs of articles.

2 Suggestive Complementarity and the ABC Model

"Complementarity", as defined above, is only suggestive, rather than compelling because scientific arguments expressed in natural language seldom lend themselves to logical description, largely because the background knowledge necessary for transforming the text of an article into a logical statement is almost always missing and typically taken for granted.

However, many scientific arguments are expressed as an association between two or more entities – such as substances and diseases in the biomedical literature. The idea of combining two entities is useful in providing a structured example of complementarity. One article might argue that term A is associated with B, and a second article that B is associated with C, in the absence of any explicit published claim that A may be related to C. This structure resembles a syllogism, but "association" and "relatedness" are not transitive so one should not be misled by the resemblance. I shall try to show that it is nonetheless useful for explanatory purposes. An AC relationship under the circumstance given would be implicitly suggested and so worth thinking about to any reader who understands both articles (A, C), a key point being the word "suggestive". Assuming that the two articles have no authors in common, it is also of interest to notice that the suggestion of an A–C relationship is unintended by the authors of either A or C who may not even have been aware of each other's work.

The ABC model, even though overly simple, is sufficiently rich to serve as a useful point of departure and as a vehicle for an organized approach to defining literature structures that have a good chance of being relevant to LBD. Moreover, the A to B to C structure can be described also by the algebra of sets, wherein we consider the set of all articles containing term A and similarly for B and for C. AB and BC are then defined in terms of set intersections. To form and combine sets of articles is the function of the core search commands for the major bibliographic databases that provide routes of intellectual access to the literature of science.

Gardner-Medwin, an eminent biomedical researcher, presciently observed in 1981: "In these days of library computers it is possible to search the literature for papers linking two or more keywords. If one were to pick out the following associations neuroglia–potassium; potassium–spreading depression; spreading depression–migraine, one would make quite an impressive collection. Try to link neuroglia with migraine however, and there would be little to show. The aim of this paper is to explore the three associations set out above" [3, 7].

Gardner-Medwin proposed and executed a core idea of LBD in a single paper with the help of a computer database search, but otherwise without benefit of computer assistance. This approach can be seen as the ABC model extended to ABCD, and was published 4 years before the work on LBD was initiated in information science, where it was called "undiscovered public knowledge" [6]. Even though the Gardner-Medwin article received about 100 citations (up to April 2007), mostly on spreading cortical depression (a neurological brain phenomenon), only six articles turned up in a Medline search on neuroglia 'AND' migraine (in the title or abstract or as medical subject headings), none earlier than 1981. Three of these cited Gardner-Medwin. Unlike the spectacular impact of the 1903 Sutton synthesis, the Gardner-Medwin hypothesis does not appear to have stimulated much further research on neuroglia and migraine, even though the intermediate steps of the connection were argued in depth, frequently cited, and well-researched. I could find no further published work by Gardner-Medwin along the line of ABC-type connections.

3 What People "Know" versus Recorded Knowledge

It is important, in the context of what is meant by "novelty" to distinguish between what people "know" (or think they know), and what is published. Literature-based discovery is concerned not with state of mind but rather with the state of the public record. It follows that the novelty of any implicit discovery hinges not on whether one or more scientists previously knew about it, but only on whether it had been previously made explicit in published form.

The journal article is one of the most important inventions supporting the infrastructure of modern science, dating from the mid-eighteenth century [10, 11]. Its function is to represent a small fragment of science, relatively short and to the point, that can then serve as a "building block" available for public use in a communal effort to construct the edifice of scientific knowledge, a process in which the blocks themselves may evolve into more mature forms and interact with their neighbors to form literatures addressed to common problems.

The size of the recorded knowledge base is far beyond human capacity to assimilate, even with the division of labor that specialization makes possible. And to include implicit knowledge based on connections increases the disparity enormously. Concerning human capacity, there is perhaps one exceptional case that has been reported:

> He is the master of Balliol College What he doesn't know just isn't knowledge [1, p. 190].

4 Fragmentation of Science

The concept of LBD arises from and depends on three essential and interlocking aspects of recorded scientific knowledge – its immense size, an attempt to cope with size by specialization, and the resulting inevitable fragmentation of science into insular communities.

Specialization in science began along with the scientific journal. The patterns of communication, particularly in citation practices, are difficult to analyze prior to the era of Eugene Garfield and the citation indexes, but manual techniques with limited objectives are not infeasible. Hybridity and cytology between 1866 and 1900, the period during which Mendel's paper was reputedly neglected, is worth a closer look for our purposes.

Whether in fact the Mendel paper was neglected and, if so, why, has occasioned much published debate, but, more to the point for LBD is the paucity of published citations by cytologists to any of the hybridity literature (including Mendel) and vice versa, even though the two fields did share a common interest in the problem of hereditary transmission, at least after 1881. I was able to find, after substantial manual searching, only a few isolated examples of cross citations between the two fields, but these did not lead to significant or ongoing interaction. Judging from the citation pattern, both cytologists and hybridists seemed to be fully occupied within their own specialties, no doubt because that is why specialties developed in the first place. Sutton's breakthrough of the cytology–hybridity boundary in 1903 seemed to be virtually unprecedented.

Most of the LBD work to date has been based on the literature of biology and medicine, perhaps because the biological world is so richly interconnected. There are many scientific bibliographic databases, but the largest two in biology and medicine illustrate the immense size of the literature today, with about 16 million articles covered by Medline and 18 million by BIOSIS (Biological Abstracts) (with substantial overlap of the two databases). Both of these databases are well-organized, indexed in depth, and associated with powerful, flexible search languages. They are the preeminent routes of access to the recorded knowledge of biology and medicine. The size of this vast literature necessarily shapes the nature of problems that LBD addresses.

Fragmentation is also manifest in the growth of the published record. Specialties do not tend to grow so large as to be unmanageable; prior to that point, subspecialties are formed spontaneously. Subspecialties therefore proliferate while maintaining a more or less limited maximum size of each. The literature of science cannot grow faster than the communities that produce it, but not so with connections. Implicit connections between subspecialties grow combinatorially. LBD is challenged more by a connection explosion than by an information explosion.

5 A Problem-Oriented Approach

The various approaches to research on LBD involve in one way or another some combination of human and machine procedures. Here, in order to bring into focus underlying principles, I envision LBD primarily as a human function, but in need of computer assistance for individual biomedical researchers.

A reasonable start for individual users of an LBD system is to define a problem in their own field of research and on that basis design a customized approach appropriate to solving the problem. The creation of relevant sub-literatures – principally by conducting searches using bibliographic databases, as did Gardner-Medwin, – is of great, perhaps overriding, importance to defining problems of manageable size.

A distinction between closed vs open ended searching is relevant in this context. Any LBD search that does not begin by clearly specifying a problem can be doubly open-ended, having, like the universe, neither a beginning nor an end [1, pp. 169–171]. Wishing to avoid questions of either cosmology or theology, I prefer to assume that one always starts with a user-defined problem that anchors the beginning of a search. The terminus is then open or closed. The open terminus often may be decomposed into multiple termini defined by a list of candidate terms suggested by either a human or a computer procedure. Any single choice from the list then characterizes a closed-end search, which is necessarily a hypothesis, not an established or confirmed finding. It remains to be tested in the laboratory, clinic, or other contexts in the real world, in the usual manner of scientific investigation.

The approach described above is individualized in that it envisions that LBD serves, and is used by, a subject specialist (e.g., a biomedical scientist) engaged in research. This approach encourages a focus on what can be done now to produce scientifically acceptable and useful results. Individualized small scale trial-and-error procedures are characterized by many dead ends and a few promising paths. We can learn from both failures and successes to develop requirements and techniques for future systems. Such an approach based on dispersed knowledge and exploratory searching is conducive to evolutionary improvement.

6 Complementary but Disjoint Structures in the Literature of Science

To determine whether supposedly new information seen in a pair of complementary articles has been published explicitly elsewhere – i.e., is not really new in terms of the state of the published record, requires a thorough literature search that may be far from a straightforward exercise.

The concept of novelty is domain dependent [8]. If we were to choose the world-wide domain of all recorded knowledge, it is impossible to prove that something is novel – i.e. does not exist elsewhere. Information retrieval is, in essence, an incomplete and uncertain process [2, 6, p. 113]. Yet, for all practical purposes a limitation to the major bibliographic and citation databases, and a high-recall search, would seem to be a reasonable basis for determining whether a connection is new to the published record, at least until proved otherwise.

The definition of complementarity can in an obvious way be extended from a pair of articles to a pair of sets of articles with each set characterized by substantially the same scientific argument. The question of whether the two sets intersect is then crucial. The new information that one might hope to gain from bringing together complementary individual articles may well already be contained in any overlapping set. In short, two complementary sets that have any substantial number of articles in common are probably of little interest for LBD.

Moreover it would be unusual for two sets of articles that cite each other extensively to be disjoint – i.e. have no articles in common, so for practical purposes it is reasonable and easier to determine the intersection of the two sets, and then only in the case of small or null intersections, check also for any citations from one literature to the other. In this context, normally one would expect two disjoint clusters to be unrelated and not complementary, and two complementary clusters to overlap extensively.

The foregoing argument suggests that two sets of articles that are complementary but disjoint (CBD)would represent an unusual structure – but it is just such a structure that commands the highest interest for LBD and is or should be the prime focus of LBD research, because the implicit results of complementary relationships that can be seen or deduced are probably undocumented and hence novel. They are likely also to be unknown and unintended [8].

The concept of disjoint, as used in CBD, is an idealization not to be taken rigidly. If relatively few articles are within the intersection of two much larger sets (say A, C), few enough so that it is not too difficult to directly examine each one to assess whether it represents a biologically meaningful connection between A and C, then for practical purposes A and C are disjoint. For any intersection paper that does represent a valid connection, the citation pattern can reveal whether or not it has been neglected, as may have been the case for Mendel's paper. LBD then might play a key role in strengthening and updating the literature-based connections (by analyzing more B-terms), and so calling attention to any neglected discovery that it might represent. I have given an example of such literature-based resurrection in a recent publication [9, pp. 1088, 1091].

7 Summing Up

I have suggested one way of thinking about literature-based discovery, stressing the point that understanding goals, problems, and concepts should precede consideration of how computers can be used to best advantage. LBD originates with the scientist as user defining a problem of interest and then examining combinations of articles that together suggest new hypotheses not apparent in the separate articles. These combinations are to be found in complementary but disjoint (CBD)literatures, the process of recognizing complementarity depending on human ingenuity. CBD literatures are formed by searching the major bibliographic databases, beginning with a user-defined problem and appropriate search strategies. The goal of an LBD system should be to stimulate human creativity in order to produce a plausible and testable hypothesis stated in a form suitable for publication in the subject field studied, where it is then open to testing, criticism, review, and stimulation of further research.

7.1 Postscript: A Warning About Consequences

In connection with a procedure very like LBD, a serious adverse effect has been predicted:

> "...some day the piecing together of dissociated knowledge will open up such terrifying vistas of reality,...that we shall either go mad from the revelation or flee from the deadly light into the peace and safety of a new dark age." – Lovecraft [1,4].

Acknowledgements I thank Neil Smalheiser for valuable suggestions.

References

1. H.P. Barrow. *Impossibilities, The Limits of Science and the Science of Limits*. Oxford University Press, Oxford, 1998
2. D. Blair. *Wittgenstein, Language and Information: Back to the Rough Ground*. Springer, Berlin Heidelberg New York, 2006. [Part III]
3. A.R. Gardner-Medwin. Possible roles of vertebrate neuroglia in potassium dynamics, spreading depression, and migraine. *Journal of Experimental Biology*, 95:111–127, 1981
4. H.P. Lovecraft. The Call of Cthulhu. In S.T. Joshi, editor. The Call of Cthulhu and Other Weird Stories. Penguin Books Ltd., London, 1999
5. W.S. Sutton. The chromosomes in heredity. In J.A. Peters, editor, *Classic Papers in Genetics*, pp. 27–41. Prentice Hall, Englewood Cliffs, NJ, 1959
6. D.R. Swanson. Undiscovered public knowledge. *Library Quarterly*, 56:103–118, 1986
7. D.R. Swanson. Migraine and magnesium: eleven neglected connections. *Perspectives in Biology and Medicine*, 31(4):526–557, 1988
8. D.R. Swanson. Complementary structures in disjoint science literatures. In A. Bookstein, Y. Chiaramella, G. Salton, and V.V. Raghavan, editors, *Proceedings of the 14th Annual ACM Conference of Research and Development in Information Retrieval (SIGIR'91)*, pp. 280–289. ACM, New York, 1991

9. D.R. Swanson. Atrial fibrillation in athletes: implicit literature-based connections suggest that overtraining and subsequent inflammation may be a contributory mechanism. *Medical Hypotheses*, 66:1085–1092, 2006
10. J.M. Ziman. *Public Knowledge*. Cambridge University Press, Cambridge, 1968
11. J.M. Ziman. Information, communication, knowledge. *Nature*, 224:318–324, 1969

The Place of Literature-Based Discovery in Contemporary Scientific Practice

Neil R. Smalheiser and Vetle I. Torvik

Abstract In this brief essay, we consider some of the lessons that we learned from our experience working with the Arrowsmith consortium that may have implications for the field of literature-based discovery (LBD) as a whole.

Keywords: Literature-based discovery · Informatics · Text mining · Hypothesis generation

1 Introduction

For the past 5 years, the Arrowsmith consortium has developed a suite of web-based informatics tools to assist biomedical investigators in making discoveries and establishing collaborations. Researchers working in multi-disciplinary neuroscience research groups have served as field testers, and feedback arising from their use of the tools in their daily work has contributed crucially to the project. We have recently described the evolution of the two-node search interface [1], discussed the role of field testers in detail [2], and described a quantitative model for ranking B-terms according to their likely relevance for linking two disparate sets of articles in a meaningful manner [3]. These and other references are available for download on the public UIC Arrowsmith website (http://arrowsmith.psych.uic.edu). Here, we would like to consider some of the lessons that we learned that may have implications for the field of literature-based discovery (LBD) as a whole.

First, what is included in the term literature-based discovery? Most authors who have used the term have referred to the so-called "one-node" or open-ended search, in which a scientific problem is represented by a set of articles (or literature) that discusses the problem, and the goal is to find some other (generally disjoint) set

N.R. Smalheiser and V.I. Torvik
UIC Psychiatric Institute MC912, 1601 W. Taylor Street, Chicago, IL 60612, USA
neils@uic.edu

P. Bruza and M. Weeber (eds.), *Literature-based Discovery*,
Springer Series in Information Science and Knowledge Management 15.

of articles containing information that can contribute to the solution of the problem [4–17]. The Arrowsmith consortium has focused primarily on the "two-node" search, in which a scientist wishes to find or assess links that connect two different sets of articles (again, generally disjoint and in different disciplines) [4, 18]. Don Swanson has proposed the term "undiscovered public knowledge" to refer to the overall process of assembling different bits of knowledge that are scattered across different literatures into a novel hypothesis [19]. Smalheiser has published numerous examples that fall more specifically into the category of "gap analysis" – that is, not so much proposing new solutions to an existing problem, but rather identifying new and potentially important scientific problems that no one seems to be studying or even noticing, either because they fall in the cracks between disciplines or for other sociological reasons [20–24]. Some LBD analyses consider discrete problems, e.g., dietary restriction in aging [25], whereas other analyses comprise more global analyses of entire disciplines, e.g., fullerene research [26]. Some studies make "incremental" predictions such as expanding the list of diseases that can be treated by a given drug [14], whereas some analyses find connections between disparate disciplines (e.g., gene therapy vs. bioterrorism) that have few articles or practitioners in common [23].

Regardless of the particular type or flavor of LBD that is pursued by different individuals, all share a more ambitious agenda than simply to extract or process the information present in a given text. If much of the research in "text mining" seeks to identify relationships that are explicitly stated, then LBD goes further to identify relationships that are implicitly stated – and not within a single document, but across multiple documents. This is a form of "data mining," but most data mining seeks to identify valid relationships within the data, whether or not they have ever been observed previously. In contrast, LBD practitioners have tended to focus on a search for relationships that are entirely novel, never noticed and perhaps never even speculated upon by scientists previously. Thus, the LBD field has set its sights on a very high, perhaps an impossibly high standard: true, novel, un-noticed, non-trivial (and generally cross-disciplinary) scientific discoveries.

2 A Case Study

We recently published a bioinformatics analysis predicting that certain genomic repeat elements within human mRNAs, the so-called MIR/LINE-2 repeats, are likely to serve as targets of the small trans-acting noncoding antisense RNA family known as microRNAs [27]. This raised the question whether other repeat elements may also serve as microRNA targets. Because Alu elements are the most common repeats expressed within mRNAs, we focused on these and found that a family of microRNAs do, indeed, appear to target Alu-containing mRNA transcripts [28]. To look for other types of biological relationship(s) we carried out a two-node search between microRNAs and Alu: The microRNA literature consisted of 970 articles, Alu had 2,945, and the intersection was empty (Fig. 1). A total of 1,428 title words

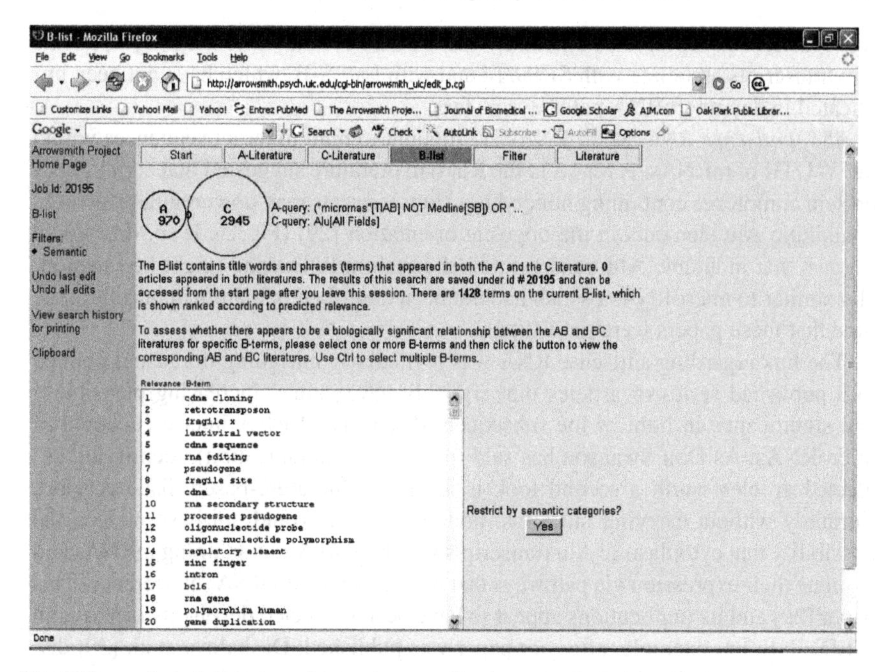

Fig. 1 Screenshot of the two-node search output for the example "microRNA vs. Alu", showing the most highly ranked B-terms

and phrases were in common to the two literatures (B-terms), and these were ranked according to a quantitative model that predicts the terms that are most likely to represent meaningful links across literatures [3]. We examined the top-ranked 100 B-terms, of which the following terms warrant discussion here:

#6, *RNA editing*. Alu repeats are highly edited, particularly within introns of unprocessed mRNAs. As well, some microRNA precursors have been shown to be edited. Finally, extensive RNA editing of a transcript inhibits its ability to be degraded via RNA interference, a pathway of RNA control that overlaps with the microRNA pathway.

#10, *RNA secondary structure*. MicroRNA precursors are characterized by a distinctive hairpin stem-loop structure. Alu repeats also express one or two hairpin loops. This raises the possibility, for example, that they might both bind proteins that recognize hairpin structures.

#25, *differentiation HL-60 cell*. It has been shown that certain microRNAs change their expression during differentiation of HL-60 cells. Separately, a set of transcripts that show significant changes in their subcellular localization and translation during differentiation were found to contain Alu sequences. Could this be a sign that certain microRNAs are targeting Alu sequences within these transcripts?

#33, *RNA binding protein*. MicroRNAs have been reported to associate with FMRP, a RNA binding protein that has an important role in synaptic plasticity. As well, both the Alu-derived small RNA BC200, and the tRNA-derived small RNA

BC1, have been reported to associate with FMRP. On the other hand, cytoplasmic Alu transcripts associate with SRP9/14 and with La/SS-B, which have not been implicated in any microRNA pathways so far.

#54, *antisense RNA*. MicroRNAs are thought to bind to target sequences within the 3′-UTR of mRNAs. A report in the Russian literature suggested that certain riboprotein complexes containing noncoding Alu transcripts may downregulate mRNAs containing Alu elements in the opposite orientation [29] (Fig. 2). If so, this would suggest that inducible Alu transcripts bind certain mRNAs and might be functionally similar to microRNAs (which had not been described in mammalian cells at the time that these papers were published).

The link regarding antisense RNA was particularly intriguing because it pointed to a published series of articles that arguably have gained increasing plausibility and significance in light of the subsequent discovery of RNA interference and of microRNAs. As Don Swanson has said (personal communication), identifying neglected articles worth a second look is a kind of literature-based discovery too! Certainly without carrying out a two-node search we would not have noticed the possibility that cytoplasmic Alu transcripts may bind to Alu-containing mRNAs and regulate their expression via pathways that may be related to RNA interference. This paper [29] and its implications appear to have been neglected despite its indexing in MEDLINE. For example, after our paper was published, Daskalova et al. [30] discussed the possibility that cytoplasmic Alu transcripts may bind to Alu-containing mRNAs, without citing any prior literature on this question.

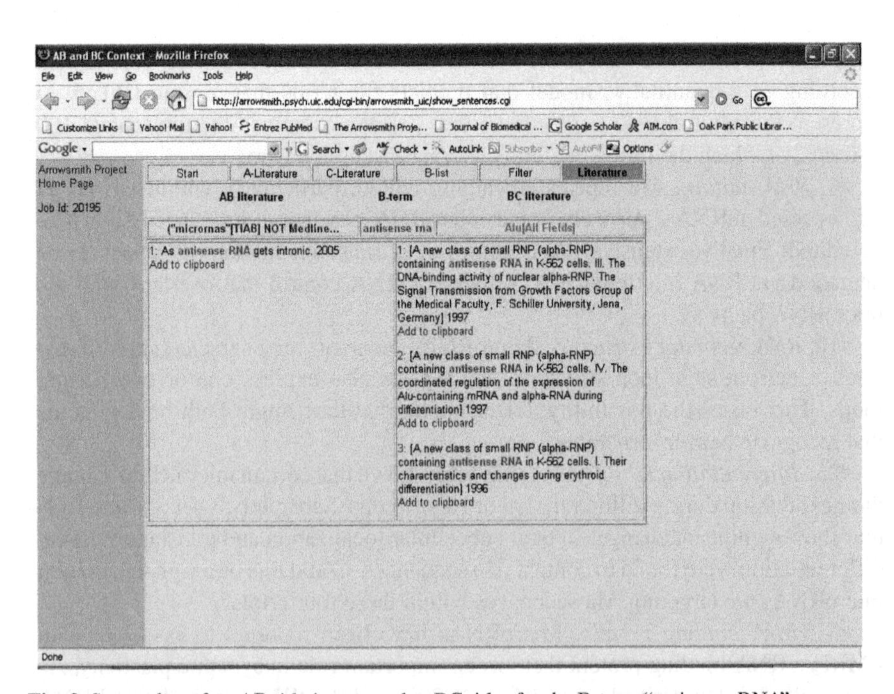

Fig. 2 Screenshot of an AB title juxtaposed to BC titles for the B-term "antisense RNA"

What is the next step? Don Swanson has proposed that a LBD finding can be considered at least partially successful if it leads to publication of a hypothesis paper in the peer-reviewed literature [31]. In the past, we have published individual LBD predictions as short notes in biomedical journals (see references listed in http://arrowsmith.psych.uic.edu). After about a dozen examples had been published, we felt that we had made the point sufficiently that Arrowsmith does assist investigators in generating and assessing hypotheses (another dozen examples of LBD findings were included in a recent description of Arrowsmith field tester behavior [2]). However, publishing papers is not a speedy, and not necessarily an efficient, mechanism for motivating other scientists to test a given hypothesis. For two hypotheses we attempted to alert investigators directly via email, but have either received no reply or short, rather dismissive responses. In principle we could arrange to test the "cytoplasmic Alu transcripts regulating mRNAs" hypothesis ourselves in the laboratory, but these experiments are neither covered by existing grants nor would obviously lead to the writing of any new grant proposals. Despite being plausible, the hypothesis remains an Orphan in search of a Daddy Warbucks.

The moral? One should not assess the value of a LBD prediction according to whether it is tested experimentally, since many pragmatic and (from the LBD viewpoint) irrelevant factors affect whether one is able to carry out an experimental test. As well, LBD is based on an analysis of the structure of the scientific literature, which reflects human activity and does NOT necessarily reflect the structure of nature – so for better or worse, one should not assess the value of LBD predictions according to whether they eventually turn out to be true after all.

3 Re-defining Success in LBD

This case is not necessarily a tragedy: After all, most hypotheses do not survive scientific scrutiny, and most are not even important enough to test at all. Yet the LBD field has defined its own success in terms of whether the hypotheses generated by LBD searches are not only truly novel and significant, but whether they have been tested by others, and whether they were validated experimentally. Given that most people who are involved with LBD do not have laboratory facilities available to test their own hypotheses, this definition of success is almost impossibly high to fulfill, and orders of magnitude beyond what is expected for any traditional search engine or IR strategy.

Let's look again at the case study in terms of what it did right: This two-node search was natural to formulate in the course of ongoing studies of microRNAs and Alu elements. It took seconds to enter the query, less than a minute to return the ranked B-list, and about a half-hour to examine 100 top ranked B-terms for titles and abstracts of the interesting papers. Thus, it did not require a large commitment of time and energy to examine, and could readily be integrated into normal workflow. The search returned non-trivial links between microRNAs and Alu elements. Although we were aware of some of these links already (and were discussed in [28]), a different person doing this search might have learned new information. Finally, the search readily identified testable and truly inter-disciplinary links between two different disparate literatures, i.e., between the microRNA field and the Alu field.

This suggests that the scope of LBD might be expanded to embrace the full continuum of information that can be retrieved from searches, from retrieval of explicitly stated information, to retrieval of implicit links, to truly novel hypotheses. Similarly, we have found that the actual information-seeking activities of field testers have completely blurred what (to an information scientist) are clear and fundamental distinctions between simple fact-finding, browsing a new literature, and carrying out one and two-node searches [2]. It should not be a surprise that end-users envision cultural products differently than do the developers – millions of people use Google without knowing how the search engine works, and billions of people enjoy music without knowing the rudiments of music theory. If LBD tools are to become popular as well, they need to be usable by people who do not know how they work. It is appropriate for the developers of LBD tools to focus on the procedural aspects and formal methodology of one-node and two-node search strategies, but to the end-users, LBD searches should appear to be simple extensions of simple PubMed searches. And, just as the end-users have blurred the distinctions between different types of information-seeking activities, so might the LBD field benefit from integrating LBD tools with other informatics resources, so that LBD comprises only one part of a larger multi-purpose tool kit. From this standpoint, success is not measured in terms of number of discoveries made, but in how many end-users utilize a given tool and how often.

4 Gold Standards

Within the community of LBD tool developers, perhaps the biggest stumbling-block to progress has been the lack of an adequate corpus of validated searches that can be utilized as gold standards. Among one-node searches, only two examples have been employed by other groups as gold standards: the Swanson studies of magnesium and migraine [32] and fish oil and Raynaud's phenomenon [33]. We initially felt that it would be impossible to create gold standards in the case of two-node searches, since given a single query (a single pair of literatures), different users might be looking for entirely different types of information. However, in the course of analyzing the two-node searches conducted by field testers, we realized two things: First, once we no longer insisted that LBD searches must predict entirely novel, untested hypotheses, it was relatively easy to ask field testers to score B-terms as relevant or non-relevant. For example, given two literatures concerning specific diseases, we could ask them to identify B-terms that correspond to surgical interventions that are performed in both diseases [2, 3]. Second, we found that B-terms that were marked as useful, interesting or relevant shared certain generic features across many different searches that distinguished them from terms that were marked as non-relevant. Thus, we were able to create manually a corpus of (currently six) diverse gold standard two-node searches, which have been employed for quantitative modeling [3] and implemented on the Arrowsmith website to rank the terms displayed on the B-list.

As well, we devised a means of creating new gold standard two-node searches and sets of "relevant" B-terms automatically, using a series of 20 templated TREC 2005 Genomics Track queries (http://trec.nist.gov/data/t14_genomics.html) asking for information describing the role(s) of a gene involved in a disease, or describing the role of a gene in a specific biological process. As part of TREC, each query was searched within a biomedical text collection representing a subset of MEDLINE, and a group of judges decided which articles were relevant to the query. We regarded the articles marked as relevant by TREC judges as "gold standards" for each query, and extracted all terms in the titles of these papers. The terms were filtered through a stoplist to remove many of the "uninteresting" terms, and the remaining terms were regarded as capturing some of the known, *explicit* information on each query. Next, we associated each query with a two-node search in which we formulated literature A = the gene name and literature C = the disease or biological process [removing any articles that mention both A and C]. The explicit title terms taken from the gold standard articles in the TREC queries serve the same function for evaluation as does the field-tester marked relevant B-terms in our own six gold standard queries [3]. We suggest that new gold standard searches can be deliberately and perhaps automatically set up for one-node searches as well. For example, in an earlier study of the potential development of viruses as biological weapons, we employed a list of viruses compiled by military experts as a gold standard [34]. One could also follow the lead of one of the Arrowsmith field testers, Ramin Homayouni, who used a set of five genes already known to be part of the reelin signaling pathway as gold standards, and applied a LBD model to a larger list of candidate genes in order to identify genes that are likely to be part of the reelin signaling pathway, even though they do not co-occur in any paper mentioning reelin [35]. This approach makes the admittedly uncertain assumption that the features of known and unknown reelin pathway genes will be similar, but this is a limitation that applies more or less to all gold standards (i.e., the assumption that new instances will be similar to the older ones, as far as their scored features).

From this perspective, it should be an easy process to generate gold standards for one-node searches, as long as two points are kept in mind. First, one must remember that LBD is attempting to model the structure of the scientific literature, not of nature, so making a list derived directly from experimental results, e.g. microarray data, does not suffice to construct a gold standard. Second, one must distinguish LBD systems that make "incremental" predictions from those that attempt to make more radical, cross-disciplinary predictions. Predicting new genes that interact with reelin is an example of the former case. Here, the LBD system merely needs to compare the features associated with a new example against a panel of known positive and negative examples, and identify those that are overall most similar to known examples. In contrast, cross-disciplinary LBD seeks to relate literatures that may appear to have little or nothing in common. The relevant measure is COMPLEMENTARITY, rather than SIMILARITY, since a particular item or concept may link two literatures meaningfully even if it is not prominent in either literature.

Re-defining success in LBD also leads us to re-assess the dichotomy that has been stated as existing between computer-generated and computer-assisted discovery tools. Certainly, we are interested in discoveries that are made by people, not by computers [4] – and yet we have found that B-terms can be automatically ranked in terms of the likelihood that one or more users will find them to be "useful" or "interesting" [3]. Evidently during data mining some nuggets can be seen to be shinier than others, and the computer can present these to the user for further inspection. Actually, the problem is not so much that computers are limited in their ability to predict new discoveries, as that individual scientists vary so widely in their interests and intuitions. It is virtually impossible for any group of scientists to reach consensus in deciding whether a truly novel hypothesis is promising and significant to follow up!

5 Concluding Remarks

If jazz is a sophisticated, intricate form of expression appreciated by the cognescenti, then LBD may be the jazz of informatics. However, jazz enthusiasts probably do not care whether the music they listen to is popular or not, whereas LBD tools were designed for working scientists and our shared goal is to make them both useful and easy to use. Our experiences with the Arrowsmith two-node search have suggested lessons that, we believe, should apply generally to other LBD projects. Most importantly, in LBD, as in jazz, we will best succeed when our different voices and instruments harmonize together.

Acknowledgements The Arrowsmith consortium is supported by NIH, and includes subcontract principal investigators Don Swanson, Maryann Martone, Ramin Homayouni, and Robert Bilder. We also thank Marc Weeber for his assistance and discussions over the years.

References

1. Smalheiser, N. R.: The Arrowsmith project: 2005 status report. In: Conference on Discovery Science 2005. Lecture Notes in Artificial Intelligence, vol. 3735, eds. A. Hoffmann, H. Motoda, and T. Scheffer, Springer, Berlin Heidelberg New York (2005) 26–43
2. Smalheiser, N. R., Torvik, V. I., Bischoff-Grethe, A., Burhans, L. B., Gabriel, M., Homayouni, R., Kashef, A., Martone, M. E., Perkins, G. A., Price, D. L., Talk, A. C., West, R.: Collaborative development of the Arrowsmith two-node search interface designed for laboratory investigators. J. Biomed. Discov. Collab. **1** (2006) 8
3. Torvik, V. I., Smalheiser, N. R.: A quantitative model for linking two disparate sets of articles in Medline. Bioinformatics **23** (2007) 1658–1665
4. Swanson, D. R., Smalheiser, N. R.: An interactive system for finding complementary literatures: a stimulus to scientific discovery. Artif. Intell. **91** (1997) 183–203
5. Bekhuis, T.: Conceptual biology, hypothesis discovery, and text mining: Swanson's legacy. Biomed. Digit. Libr. **3** (2006) 2

6. Jensen, L. J., Saric, J., Bork, P.: Literature mining for the biologist: from information retrieval to biological discovery. Nat. Rev. Genet. **7** (2006) 119–129
7. Weeber, M., Kors, J. A., Mons, B.: Online tools to support literature-based discovery in the life sciences. Brief. Bioinform. **6** (2005) 277–286
8. Hristovski, D., Peterlin, B., Mitchell, J. A., Humphrey, S. M.: Using literature-based discovery to identify disease candidate genes. Int. J. Med. Inform. **74** (2005) 289–298
9. Skeels, M. M., Henning, K., Yetisgen-Yildiz, M., Pratt, W.: Interaction Design for Literature-Based Discovery. Proceedings of the ACM International Conference on Human Factors in Computing Systems (CHI 2005), Portland, OR
10. Wren, J. D.: Extending the mutual information measure to rank inferred literature relationships. BMC Bioinformatics **5** (2004) 145
11. Srinivasan, P., Libbus, B.: Mining MEDLINE for implicit links between dietary substances and diseases. Bioinformatics **20** Suppl 1 (2004) I290–I296
12. Wren, J. D., Bekeredjian, R., Stewart, J. A., Shohet, R. V., Garner, H. R.: Knowledge discovery by automated identification and ranking of implicit relationships. Bioinformatics **20** (2004) 389–398
13. Srinivasan, P.: Text Mining: Generating hypotheses from MEDLINE. J. Am. Soc. Inf. Sci. Technol. **55** (2004) 396–413
14. Weeber, M., Vos, R., Klein, H., De Jong-Van Den Berg, L. T., Aronson, A. R., Molema, G.: Generating hypotheses by discovering implicit associations in the literature: a case report of a search for new potential therapeutic uses for thalidomide. J. Am. Med. Inform. Assoc. **10** (2003) 252–259
15. Weeber, M., Vos, R., Baayen, R. H.: Using concepts in literature-based discovery: simulating Swanson's Raynaud–fish oil and migraine–magnesium discoveries. J. Am. Soc. Inf. Sci. Technol. **52** (2001) 548–557
16. Valdes-Perez, R. E.: Principles of human-computer collaboration for knowledge discovery in science. Artif. Intell. **107** (1999) 335–346
17. Kostoff, R. N.: Science and technology innovation. Technovation **19** (1999) 593–604
18. Smalheiser, N. R., Swanson, D. R.: Using ARROWSMITH: a computer-assisted approach to formulating and assessing scientific hypotheses. Comput. Methods Programs Biomed. **57** (1998) 149–153
19. Swanson, D. R.: Undiscovered public knowledge. Libr. Q. **56** (1986) 103–118
20. Smalheiser, N. R., Swanson, D. R.: Assessing a gap in the biomedical literature: magnesium deficiency and neurologic disease. Neurosci. Res. Commun. **15** (1994) 1–9
21. Smalheiser, N. R., Manev, H., Costa, E.: RNAi and Memory: Was McConnell on the right track after all? Trends Neurosci. **24** (2001) 216–218
22. Smalheiser, N. R.: Predicting emerging technologies with the aid of text-based data mining: the micro approach. Technovation **21** (2001) 689–693
23. Swanson, D. R., Smalheiser, N. R., Bookstein, A.: Information discovery from complementary literatures: categorizing viruses as potential weapons. J. Am. Soc. Inf. Sci. Technol. **52** (2001) 797–812
24. Smalheiser, N. R.: Bath toys: a source of gastrointestinal infection. N. Engl. J. Med. **350** (2003) 521
25. Fuller, S. S., Revere, D., Bugni, P. F., Martin, G. M.: A knowledgebase system to enhance scientific discovery: Telemakus. Biomed. Digit. Libr. **1** (2004) 2
26. Kostoff, R. N., Braun, T., Schubert, A., Toothman, D. R., Humenik, J. A.: Fullerene data mining using bibliometrics and database tomography. J. Chem. Inf. Comput. Sci. **40** (2000) 19–39
27. Smalheiser, N. R., Torvik, V. I.: Mammalian microRNAs derived from genomic repeats. Trends Genet. **21** (2005) 322–326
28. Smalheiser, N. R., Torvik, V. I.: Alu elements within human mRNAs are probable microRNA targets. Trends Genet. **22** (2006) 532–536
29. Petukhova, O. A., Mittenberg, A. G., Kulichkova, V. A., Kozhukharova, I. V., Ermolaeva, I. uB., Gauze, L. N., Konstantinova, I. M.: A new class of small RNP (alpha-RNP) containing antisense RNA in K-562 cells. IV. The coordinated regulation of the expression of Alu-containing mRNA and alpha-RNA during differentiation. Ontogenez **28** (1997) 437–444

30. Daskalova, E., Baev, V., Rusinov, V., Minkov, I.: Sites and mediators of network miRNA-based regulatory interactions. Evol. Bioinform. Online **2** (2006) 99–116
31. Swanson D. R.: Intervening in the life cycles of scientific knowledge. Libr. Trends **41** (1993) 606–631
32. Swanson D. R.: Migraine and magnesium: eleven neglected connections. Perspect. Biol. Med. **31** (1998) 526–557
33. Swanson D. R.: Fish oil, Raynaud's syndrome, and undiscovered public knowledge. Perspect. Biol. Med. **30** (1996) 7–18
34. Swanson, D. R., Smalheiser, N. R., Bookstein, A.: Information discovery from complementary literatures: categorizing viruses as potential weapons. J. Am. Soc. Inf. Sci. Technol. **52** (2001) 797–812
35. Homayouni, R., Heinrich, K., Wei, L., Berry, M. W.: Gene clustering by latent semantic indexing of MEDLINE abstracts. Bioinformatics **21** (2005) 104–115

The Tip of the Iceberg: The Quest for Innovation at the Base of the Pyramid

M.D. Gordon and N.F. Awad

Abstract Much of the world in Asia, Latin America, and Africa is at an early stage of economic development similar to what the United States and other developed countries experienced many decades ago. Yet, much as their needs for hard and soft infrastructure, effective business practices, and an educated workforce parallel similar needs that underlay earlier development in the West, replicating Western development would overlook the hallmarks of the current century: widely available information and communications technology; a set of electronic linkages among the world; and a global business environment, to name just a few. Consequently, it should be possible to allow developing countries to use "leapfrog" technologies that were inconceivable decades ago to support their development. One means of identifying these opportunities is by matching traditional development needs with novel support by connecting previously unrelated literatures.

Equally interesting, the poor in many regions are compelled to seek innovative solutions that extend their resources and otherwise make their lives easier. These can include truly surprising hybrids (like washing machine – bicycles) that serve distinct local needs. Yet, these innovations have the potential to be of great value in West, either through direct commercialization or serving as a source of inspiration. These developing world innovations, too, can be linked to currently unrecognized needs or opportunities in the West by proper cross-fertilization. Again, literature-based methods may be an effective means to discover mutual benefits linking the developing and developed worlds.

M.D. Gordon
Ross School of Business, University of Michigan, USA
mdgordon@umich.edu

N.F. Awad
School of Business, Wayne State University, USA
nawad@wayne.edu

P. Bruza and M. Weeber (eds.), *Literature-based Discovery,*
Springer Series in Information Science and Knowledge Management 15.

1 The Base of the Pyramid and Innovation

The economic base (or bottom) of the pyramid represents the two to five billion people living lives at or barely above the poverty level. The description comes from imagining the world's population divided into strata according to personal income, and then layering these strata one upon the other, the lowest income at the bottom. The resulting shape would be something like a pyramid, with vast numbers of the world's population living at its base. As Prahalad and Hart [28] point out in their article, "The Fortune at the Bottom of the Pyramid," approximately four billion of the six billion people on the planet live on $4 a day or less (calculated in terms of PPP, purchasing power parity, meaning what $4 would buy in the U.S.). Of these, several billion live on less than $2 or even $1 per day. An additional 1.5 to nearly 2 billion people live on incomes above $4 daily, but still less than $20,000 annually. Finally, there are a few hundred million people living on annual incomes above $20,000. The bulk of the very poor live in India (population, 1.1 billion), China (1.3 billion), and Africa/Latin America & Caribbean (0.85 billion).

The problems facing the poor are those that too naturally flow from living on such low incomes. Health problems are magnified because of inadequate healthcare facilities or the poor's inability to afford the healthcare that is available. These medical problems include many preventable and treatable diseases, such as malaria, a debilitating and often deadly affliction for millions. Similarly, sanitation and housing are normally not of a standard to support a healthy life. Lack of adequate educational opportunities are among the other deficits the poor encounter.

Many approaches have been suggested for providing better economic opportunity for those with such little means[1]. For over a half century, governments in the developed world and institutions such as The World Bank, International Monetary Fund, and other multilateral institutions and charities have offered financial and other support. Despite the failure of these efforts to eliminate or even deeply dent poverty, ambitious new efforts for directing and administering aid have been offered. The Millennium Villages Project in Africa, championed by Jeffrey Sachs, aims to provide an integrated set of scientific and economic remedies to lift people out of poverty [23]. Lodge and Wilson [22] propose the establishment of a permanent partnership among key MNCs, aid agencies, and NGOs to help define a series of economic and development projects and the right actors to carry them out in specific parts of the world [22].

An approach that has captured the public imagination is using the capabilities of business to sell to and build businesses supporting job and business creation at the bottom of the pyramid ([28]; Prahalad 2005; Hart 2005). In aggregate the wealth of the poor is staggering, because of their sheer numbers. And they are a source of energetic, innovative energy, if only it can be tapped. As Prahalad opens his book:

[1] No one has suggested using the literature based discovery techniques for assistance. Showing how this may be done is at the heart of this paper.

> If we stop thinking of the poor as victims or as a burden and start recognizing them as resilient and creative entrepreneurs and value-conscious consumers, a whole new world of opportunity will open up. Four billion poor can be the engine of the next round of global trade and prosperity... [and] a source of innovations.

A key point in Prahalad's thinking is occasionally overlooked or misstated. He believes that a business ecosystem, with multinational companies playing a central "nodal" role, will help unleash the entrepreneurial spirit of the poor. They will sense the opportunity to provide and improve products or services, and with an infrastructure in place, will take advantage of the opportunity and do so.

A stronger version of these sentiments is offered by Professor Anil Gupta, of the Indian Institute of Management, Ahmedabad. The rural poor represent the "tip of the (economic) iceberg": what you see hides what lies beneath – enormous potential to produce, a tremendous storehouse of innovation and creativity. Why? Simply because the poor understand their needs and have found innovative ways to meet them with the limited resources at their disposal. One of the definitions of innovation captures the essence of why the poor offer tremendous innovation potential: "(Innovation is) a mindset, a pervasive attitude, or a way of thinking focused beyond the present into the future" [20].

Gupta has established social systems for scouring the rural countryside to identify individuals with innovative, generative capacity. By means of these shodh yatras, or journeys of discovery, he has identified tens of thousands of rural innovations and applications of indigenous knowledge. Innovations range in size, scope, application, and degree of sophistication. Many involve applications of local knowledge for veterinary or human purposes, crop protection etc. Others involve technologies for heating, refrigeration, communication, and farming.

Interestingly, the bulk of innovation occurs in one specific rural area of India, the state of Gujarat. With limited local assets and access to external resources (finance, technology, information), rural communities face significant livelihood challenges. Communities derive goods and services from five types of assets: (1) natural, (2) social, (3) human, (4) physical, and (5) financial [7]. In rural communities, due to a shortage of other assets, the key resource available for the poor is social capital; communities must work together for the sake of their livelihood [11, 12, 19, 30, 37]. In vulnerable and marginal communities, the need for innovation is imperative [8]. The locus of rural innovation in India, therefore, occurs in an area where significant social capital has been developed, both within the community and, to a degree, with the surrounding business community that represents outside innovations. A handful of products have been patented, beginning their climb from local to potentially broader application, with the resulting possibilities of generating income for their inventor and local job creation as well. A small number of institutions in India developed by Gupta and others support the conversion of innovation into marketable product.

A stylized flow diagram captures some of the necessary transformations that must occur for an innovation to reach its potential impact. The flow diagram is aligned with source-based stage models of innovation that flow from idea inception to final product, with the person who initiates the new idea, the innovator, being the source [1, 17, 36, 40]. Based on a legitimate local need, an idea arises to address it. (Indeed a network of innovators may incorporate suggestions from each other, adopt each others' ideas, and otherwise support each other, therefore utilizing their social capital.) At the stage of invention, the right (local resources) must be obtained along, possibly, with necessary tooling to produce a series of ever-refined, ever-informative prototypes. In certain innovations, such as using Teflon coatings for clay pots, experiments may rapidly determine the best course of action (the best means of application of liquid Teflon and the amount to apply). More complicated innovations may require months or years of experimentation and many widely varying product designs. Naturally, user trials and acceptance and other forms of technical feedback are fed into the prototyping process. Once ready for use, the innovation faces the challenges of the market. These certainly include but are not limited to considerations such as manufacturing, distribution, marketing, etc. Legal, policy, and other societal considerations may also be key in assuring commercial success.

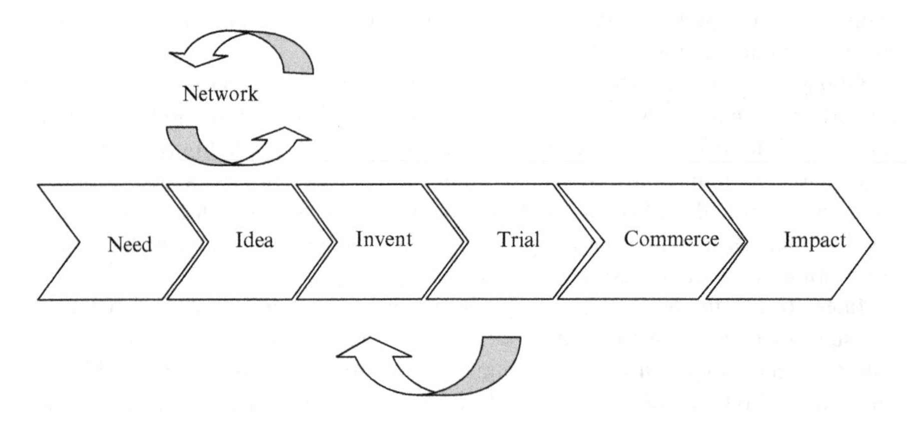

Innovations based on using local plants for medicinal or other beneficial means is a bit different. There have long been local uses for plants and herbs. But skepticism, secrecy, and indifference have sometimes caused the usage of extremely effective plants and herbs to be overlooked as a means of providing benefits to large numbers of people. For instance, one of the drugs currently part of the modern armamentarium for combatting malaria is artemisinin. The Chinese have used sweet wormwood (artemisia annua L.), from which artemisinin is extracted, for at least a century for its medicinal uses (including to combat malaria the last several decades). Ninety percent effective in combination with other drugs, and endorsed by the World Trade Organization, the beneficial effects of wormwood escaped the attention of the West until recently. Now, intensive efforts are under way to produce and extract sufficient quantities using modern practices to provide artemisinin based therapy widely. Such a flow of awareness is aligned with user-based stage models of innovation, which are based on the perspective of the user, rather than the innovator, and follow the innovation process from the user's awareness to the incorporation of the innovation into the user's behavioral routines [3,25,36,39]. Increasingly, local innovations from rural communities are being acknowledged by a global set of users. Traditional (indigenous) knowledge is being recognized for an increasing number of uses in addition to human and animal healthcare. Among them are: supporting pest control, crop diversity, soil conservation, and water management.

In the cases of both the invention of physical goods at the base of the pyramid (mechanical devices, electronic devices, etc.) and the application of indigenous knowledge, there is great potential for finding broader acceptance or even new applications. One possible means for doing so is through connections to companies and organizations in the West. A Western business, for instance, can provide support at any stage of the value chain. Conversely, the new ideas flowing from the innovators may provide opportunities for the Western companies to fulfill unanswered needs of their customers, or to develop novel offerings based on inspiration from rural innovators. A resulting innovation may consequently be improved in terms of the features it contains, the materials it uses, or its manufacturability. The resulting price: performance ratios can be dramatically altered, as in these cases in India: making and fitting of prosthetic limbs and performing eye and heart surgeries, where the cost advantages are 40, 50, and 200 *times*, respectively. For performance comparable to the highest Western standards. Western firms can be instrumental in helping translate these advantages to broad markets, worldwide.

We can begin to understand the possible mutual benefits of Indian innovators working with Western institutions by looking at how innovations may fail to become widely adopted commercially. An innovation that truly meets a local need may suffer in its journey towards commercial application at different stages. Funding, technical expertise, or certain design principles may be lacking during the creation of a series of prototypes. Initial trials may be thwarted by the lack of funds or other resources for manufacturing, distributing, or testing. Trials may indicate the need for product enhancements, for which additional resources are required but lacking. Successful local trials that indicate market acceptance benefit from expertise in and possible partnerships in sales, marketing, distribution, licensing, and securing appropriate intellectual property rights (an issue that is especially crucial where local, indigenous knowledge may be threatened by outside commercial interests). Again, resources in the form of capital, knowledge, and overall support for business execution will all be required for an innovation to gain significant traction.

Using networks of resources to foster the innovation process has been discussed in various literatures, including: innovation [13,31], sociology of science [21], and sociology of economic institutions [26]. Networks facilitate the ability to transmit and learn new knowledge and skills [26]. Accordingly, innovation is increasingly fostered through networks of learning that involve various entities and organizations [27]. One such example of a successful network of innovation is open-source software development [5,41]. However, developing a network does not relieve the need for the local innovation participants to be capable and innovative, nor does it eliminate the importance of Western institutions providing necessary resources for execution. Rather, it merges the two [2,24]. We examine several innovation networks in the context of the base of the pyramid and literature-based discovery towards the end of the paper.

2 Disruptive Technologies

In the mid twentieth century, Schumpeter [32] coined the phrased "creative destruction." With parallels to biological adaptation, creative destruction seeks ever better adapted forms of business and industry, at once creating new value but destroying incumbent firms and industries. Firms seeking to grow their markets can be trapped by the expectations of the their customers, their own relentless drive to improve their existing products (though in incremental ways), and management and accounting systems that are ill-suited to establishing and funding market-place experiments. Small, hungry firms, on the other hand, can much more easily introduce products that people in the developing world would hunger for but that would accurately be perceived as sub-standard in the developed world. For instance, Christensen and Hart [9] and Christensen et al. [10] talk about extremely low cost ($3,000) mini-vehicles in China (GM participated in this joint venture) and other "disruptive technologies" like stripped down microwaves. These, though, still fail to address markets at the true bottom of the pyramid. In contrast, disruptive communications technologies like n-Logue's wireless, broadband (in India); Grameen's (Bangladesh) disruptive service model for providing cellular service; and solar photovoltaic, wind, fuel-cell, and micro-turbine power generation (across the developing world) currently provide solutions to the real problems and needs of the poorest of the poor. It is easy to understand that these technologies *only* address the needs of the poor. Who in the West, for instance, wants intermittent, relatively expensive, limited electrical power? Yet, as such technology takes root in the developing world – where it is truly welcome since it far surpasses alternatives on these dimensions – its quality, performance, and price will all improve. Such perfected disruptions make them formidable candidates for moving cross- or up-market to compete with long-entrenched technologies that can begin to appear outdated.

Writing in *Seeing Differently: Insights on Innovation*, Bower and Christensen [4] summarize the qualities of disruptive innovations. The technologies do *not* meet current Western needs along one or several important performance dimensions. But, over a relatively short period of time, as they are widely trialed and adopted, they will dramatically improve to the point where they can successfully invade mature, existing markets. For Western firms not to be caught off guard and miss out on these advances, they advise them to avoid their traditional channels in gauging these new markets and, rather, to let other, nimbler organizations conduct experiments but to monitor them closely.

The development of disruptive technologies is supported by the trend towards democratizing [41] and distributing innovation. Web sites such as digitaldividen.org, nextbillion.net, and thinkcylce.com foster open access and collaboration directed at innovation. Such communities are sponsored by non-profits, academic institutions, and for-profits. Each uses information-technology to support social connectedness, collaboration, and innovation across various communities, including those communities that are underserved. We suggest supporting these innovation communities in a new way through literature-based discovery.

3 Application of Literature Based Discovery

With proper linkages, the ingenuity, creativity, and resourcefulness of rural innovators can be joined with the business, financial, and engineering muscle of the West. Both sides can benefit. Though other means of forging such linkages are, of course, possible, we suggest that literature based discovery may play a productive role in forging them.

A few years ago, Gordon et al. [15] discussed several modifications to literature based discovery as it is currently practiced. To begin with we suggested that literatures other than MEDLINE were appropriate starting points, the advantages of the uniformity and careful indexing of MEDLINE notwithstanding. The same (over-) specialization that makes it difficult to connect ideas within medicine suggests that there are missed connections and missed opportunities in other areas of application. These likely occur far more often than we suspect, primarily because it is hard to know what you're not seeing when you're not seeing it. Historically, the widespread adoption and uses of the telephone, movie projector, mainframe computer, and the Internet were all overlooked or missed entirely by individuals and companies that should have been in position to know. For example, in the biomedical arena, Richard DiMarchi, Vice President for Endocrine Research at Eli Lilly and Company, emphasized that the biggest mistake his company could make in managing research alliances was to treat them as "one-offs" – independent relationships pursued separately [27] – rather than see the continuing potential for innovation.

Gordon et al. went on to suggest that the "direction" of literature-based discovery could be turned around. Vos [42] argued that drugs produce a set of effects, some wanted and others initially viewed as negative "side effects." Yet, these side effects can produce blockbuster drugs. The drug minoxidil, developed for hypertension, became the baldness drug Rogaine. Sildenafil citrate was an unpromising drug for alleviating chest pains that has had far more success as Viagra. Weeber et al. [43] proposed a literature based discovery support architecture based on Swanson's [34] pioneering efforts that could be used for matching the "side effects" of drugs with conditions where these effects would be wanted.

We suggested inverting the disease-cure trajectory that Swanson had initially proposed and applying it to find new applications for technology in any discipline, not just medicine. In a series of experiments on the World Wide Web, we demonstrated how what we called *extension* was possible in the area of computer science. The figures below (adapted from Gordon et al. (2002)) show the flow of "traditional" literature based discovery and extension, respectively. It is interesting to note that "traditional" literature based discovery is aligned with source-based models of innovation, which start with an innovator addressing a problem. Literature-based discovery using extension corresponds with user-based models of innovation, which start with user awareness of a new application or innovation.

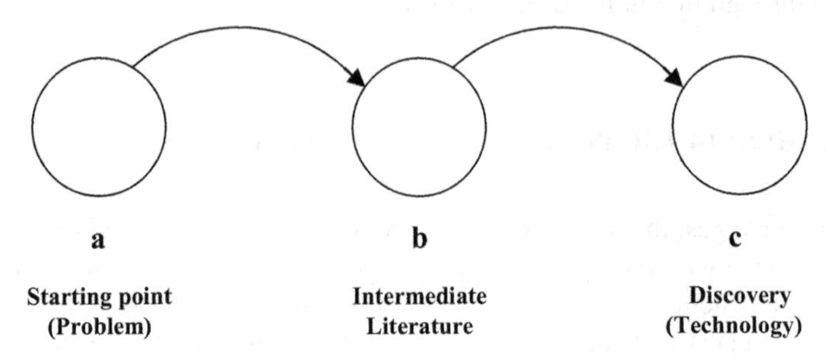

Flow of Traditional Literature-Based Discovery

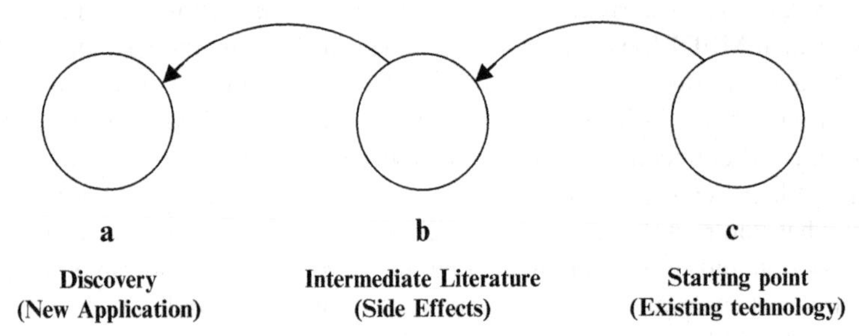

Flow of Extension in Literature-Based Discovery

We made two other observations in Gordon et al. [15] that we also feel apply to linking the West and rural innovators in India. One, the process of analyzing a literature can be abbreviated. We suggested that an intermediate, B, literature, can be chosen without first analyzing the starting literature. It could be done on the basis of a researcher's prior knowledge or even her hunch about which subsequent analysis would be most profitable. We will take a position in the current paper that is similar to this. Two, absolute novelty is not the only objective of literature-based discovery. Finding connections that are new to the *investigator* can be equally important – especially in commercial settings. That position, too, will apply to the current work.

Let us now consider in a conceptual sense how literature-based discovery might link Western, commercial interests with innovations at the base of the pyramid. We will suggest that West–South linkages may serve both as a source of new ideas for Western firms and as a means by which the skills and talents in the West provide advantage for the developing world. Already, Western companies realize that to compete they must innovate; and to innovate they must begin to throw out the rule book. Teece [35] argued that the rise in cooperative innovations has shifted the innovation process to make it more distributed, such that fewer organizations are truly able to innovate by operating on their own. Large companies have begun to form internal divisions with responsibilities for identifying and selecting appropriate innovations from around the world. Procter & Gamble incorporates a strategy it calls "Connect and Develop," through which it hopes that at least half of its new products are invented outside the company. The concepts from the open-source software development community guide P&G's and other firms' efforts: There are many talented people around the world, with lots of knowledge and complementary aims and ambitions. It is foolish to think you can top this by relying exclusively on the resources within your own firm.

To identify these new ideas, companies send innovation scouts around the world, typically to research labs, start-ups, scientific conferences and other well-understood avenues for exploring innovation. Yet the base of the pyramid offers vast untapped potential for companies to identify more disruptive technologies. Databases established to support rural innovation catalog over 30,000 herbs and plants used locally in human and veterinary medicine, pest control, land conservation, etc. Mechanical and other man-made rural innovations are catalogued separately. On the one hand, Western firms with access to these databases can search for ideas that extend their portfolios. Such databases are examples of objects that forge connections among actors collaborating across an innovation network [6, 21]. Objects that mediate the relationships among actors are important for fostering innovation [33]. They include standards [14], platforms [18], and databases [41]. Web sites supporting community are emerging as new mediation forms between objects. These websites can connect potential innovation collaborators across the globe.

The National Register of Grassroots Innovation lists as available for commercialization two novel uses for motorcycles. One transformed a popular line of motorcycles, Bullet motorcycles, into a low cost device for tilling and cultivating small farms. By temporarily converting a motorcycle into a three-wheeled device with a rear "toolbar" and making various mechanical modifications to its engine and differential, small, poor farmers have the ability to farm with greater productivity and reliability than relying on livestock to pull plows. A separate innovation involves using motorcycles for spraying fields. Using inexpensive materials, the sprayer taps the motorcycle engine to create pressure that is used to spray insecticide.

How might literature-based discovery support the discovery of new applications from base of the pyramid innovations? Let us think about literature-based discovery a moment from a conceptual perspective, overlooking the details of any algorithms used for implementation. The process has two phases: abstracting, then re-contextualizing. Using lexical statistics to highlight what Raynaud's is about is a way to view the condition more abstractly. Establishing that fish-oil accounts for many of these features re-contextualizes them. Similarly, creating a drug profile and then matching it to a new disease profile, abstracts and then re-conceptualizes the description of a drug.

As we have suggested, it is not always necessary to begin literature-based discovery by examining a set of documents (as is done, say, by accumulating all the Raynaud's documents for analysis). We may be able to abstract the situation from even a single source document.

How might a manufacturer of motorcycles take advantage of literature-based discoveries to create entirely new products, even new markets? In examining a description of the Bullet tractor, new designs might suggest themselves, and thus new products. Using pure intellect, search engines, an online thesaurus, or another type of tool for seeing conceptual relationships, any of the following physical re-designs that address various usage scenarios might suggest themselves:

- Substituting different kinds of *vehicles* for the motorcycle, including vehicles as different as handheld snow blowers, jet skis, etc.
- Finding different types of *blades*, possibly for mulching soil or helping pull lily pads from the swimming area of a beach.
- Developing other forms of *hybrids*, similarly to the way that road bicycles can swap out a wheel to become mounted exercises bikes.

But design changes need not be physical at all. For instance, the National Register of Innovation also includes an automatic spray pump for delivering insecticides. As the text of the description indicates, the spray is emitted by a special kind of sandal that creates pressure to operate a pump, thus eliminating the need for any kind of hand winding and making the delivery of insecticide convenient for someone who is walking over a small area. It is also far less hazardous than other forms of crank-operated sprayers. (See Fig. 1.) With only a little imagination and a focus on various potential new users, one can think of new uses for this innovation: a jogging shoe that keeps you cool either by spraying you with a mist or fanning you as you move; or a shoe (with a switch) that emit a loud sound (air pressure) either to scare animals (dogs while jogging; bears while hiking) or even to attract attention (lost child?).

Another innovation in the National Register is a remote cell phone-based starter that a farmer can use to turn on or off various pieces of machinery (irrigation pumps, etc.), sometimes on short notice, on a farm that can be a good hike away. Like the modified insecticide sandal, this device can be re-conceptualized. While we are beginning to hear talk of using computers to turn up your heat or turn on your lights before you come home, those images are more a matter of future than current technology. Yet, in rural India, this technology is close at hand for one who knows how to read between the lines in the literature on innovation. In rural India, this technology was produced out of necessity and, once published in the Register, the idea becomes a public good. However, determining which additional applications are the most appropriate private goods with the greatest market potential (a remote car starter that can operate at a distance far greater than similar devices that interact with a car's starter via weak radio waves?) is a matter of business discovery.

4 Discovery: West to South

To this point we have considered innovations whose genesis was with rural innovators in India, and whose broader application might be in more developed, Western markets. We now briefly consider how this flow of innovation might be reversed: using innovations in the West in contexts for which they were not originally intended. In this case we can consider Western markets through source-based models of innovation, and rural areas through user-based innovation models.

Automatic spray pump for insecticides

Background

Parbatbhai has a natural aptitude towards working with machines and creating new gadgets to make life easy. When he started earning enough to support has experiments first of all he made some modifications in the engines of 'Luna' and 'Bullet' two wheelers and learnt about tractor repairing. His next project was to develop a fuel-efficient submersible pump for drawing water from the wells. It took him three and a half years to develop the pump. In this new device he has replaced the electric motor blow with a hydraulic motor that runs on oil circulation. About 8 to 10 litres of crude oil for 24 hours is required to run the motor.

Invention of the spray pump

Parbatbhai used to spray insecticide on cotton crop in his field. He soon realized the tiresome and dangerous nature of the job of spraying of insecticides from the available pump. The pump required continuous winding of the handle, which was a very tiresome job. Along with the danger of the liquid spilling and harming the farmer was always there. Then, the cost of repairs was an additional burden on the farmer. These discomforts made him think of developing a spray pump, which would be rid of such problems.

Construction of the pump

First he made a spray pump working on the jerks and swings crated by the farmer's walking movements. When the farmer walked carrying the tank on his back, his movements gave jerks to the tank and insecticide was sprayed. But there was a practical problem. This pump was large and got very heavy. And if the tank was of a smaller size the liquid did not create adequate pressure. More over, it was costly also. Therefore it was not practically useful. Parbatbhai then chanced upon the invention quite accidentally. While he was making a pump he found that the tank was leaking. The leakage could not be located even after intense search. So he filled the tank with air using a foot pump. When the tank was full water sprayed out from the place of leakage. He got the insight he was waiting for to spray the pesticide using with air pressure! Then he got the idea of using air sandals in the place of screw pumps and was successful.

This spray pump did not need any winding of handles to spray because the sprayer had to wear a special kind of air sandals designed by Parbatbhai. These air sandals created air pressure, which got exerted on the tank and sprayed the liquid outside. This saved time, energy and labour cost.

Utility of the pump:

There are a number of advantages of this pump, which are as follows:

Present Spray Pump	Pump developed by Parbatbhai
1.Works only by winding of handle.	1. No need for wind a handle.
.2.Danger of insecticide spill.	2. Insecticide does not spill.
3.Winding of handle is a tiresome job.	3. This strain is not there in this pump.
4. Spray is formed by winding the handle.	4. Spray is formed by air pressure.
5. Needs repairing often.	5. Very little need of repairing.
6. Capacity 16 ltrs.	6. Capacity 16 Ltrs.
7. Weight 6 to 7 kgs.	7. Weight 2.5 to 3 kgs.
8. Needs replacement of washer	8. No washer at all.
9. One person can do one spray	9. One person can do two sprays.
10. One person can run only one line.	10. One person can run two separate lines.
11.Costs Less.	11. Cost is more compared to the type available in the in the market.
12. Labour cost is more.	12. Labour cost is less.
13. Needs a mechanic for repairing.	13. An ordinary farmer can do the repairing.
14. Spare parts cost more.	14. Spare parts cost less.
15. Spare parts are available in town, cities only.	15. Spare parts are available in the villages also.

Use by other farmers:

Parbatbhai's invention is not yet much known. He used this pump for the first time in his own field this year. He wishes to let his invention spread to all parts of the country.

Future Planning:

Parbatbhai has decided to get a registered trademark for the pump, and then get it patented for wide scale production. He wants to make a good quality item which rarely needs repairing and does not create problems.

Fig. 1 Entry in Indian Innovation Database on sandals that spray insecticide

Supported by different business models and having different technology formats, various "literature" databases are potentially available to help move innovations from the West to the South; these databases create interactions among knowledge originating from diverse and previously disconnected sources [38]. InnoCentive, a business begun by Eli Lilly, attempts to find "problem solvers" with solutions for "problem seekers" and facilitates the transfer of intellectual property rights. Covering areas such as chemistry, biology, and materials science, the potential of this literature for supporting the needs of the developing world is apparent. The commercial nature of the business may rule out its applicability for this purpose, however. yet2.com operates somewhat similarly, considering itself a virtual technology market for identifying, leveraging, and brokering deals surrounding intellectual assets. As part of its massive efforts to create innovation networks through its Connect + Develop program, Procter and Gamble uses yet2.com's search technology to allow others to take to market some of the 27,000 patented ideas it has but does not intend to develop. Some of these it simply donates, receiving tax benefits but no other compensation. One example of a technology that P&G holds that might support the developing world is a low-power electrolysis technology for disinfecting a water supply. They suggest the technology is scalable, kills most pathogens, and runs on a variety of power sources including batteries and solar. See Fig. 2.

Other business models are more readily applicable for making technology accessible to those in the developing world. OneWorld Health, a non-profit pharmaceutical company, finds discarded drugs that other organizations may be willing to donate for new uses in the developing world, and then takes the drug through the normal stages of drug research, screening, testing, and, ultimately, manufacturing and screening. It then arranges for manufacturing in the developing world (to produce jobs) and ensures the drug's distribution to those in need.

Electrolysis Cell Inexpensively Disinfects Water

All the world ultimately obtains its water from local, natural sources. Not even the best municipal water system in the developed world, however, is 100% effective at killing and removing water's pathogens—and many parts of the world don't have even that. This low-power electrolysis technology can disinfect a reservoir of water such as a storage tank or pool, or be placed in-line to the water supply to kill the bacteria, viruses, parasites, protozoa, molds, and spores that find their way into the water used for drinking, cooking, bathing, cleaning, and other personal uses. Scalable up or down, the technology can run on current, batteries, or even on solar power. Efficient, effective, and relatively inexpensive to manufacture and operate, the decontamination cell makes water safe.

Benefits Summary

• Disinfects water inexpensively. • Creates a dilute solution of mixed oxidants to disinfect water. • Inexpensive to manufacture. • Inexpensive to operate. • Able to be packaged in many different forms suitable for a variety of applications. more

Development Summary

Electrolysis cells have been created and produce mixed oxidants. A cell has been incorporated into a spray bottle, where it runs on two AA batteries. more

IP Summary

This technology is supported by 2 US patents. more

Fig. 2 Water disinfectant listed on yet2.com

The BiOS initiative also directly addresses the needs of the developing world. Following an open source (OS) approach in biology (Bi), the initiative supports collaborative development of innovative biological technologies, providing open access to patented and unpatented intellectual property to those traditionally shut out (such as the public sector), while still protecting commercial rights from developing new products. BiOS provides a set of tools, including a literature database providing information about both technologies and patent/IP properties.

5 New Frontiers in Literature-Based Discovery

This article has suggested that innovators in rural India have ideas that may potentially be brought to market with success if appropriate linkages with Western organizations are established. Likewise, we have suggested that technologies in the West have the potential to make significant differences in the lives of the poor. Innovation networks in both the West and in rural India serve as a source of exchanging ideas, technologies (and sometimes encouragement). These networks are virtual, with all exchange being mediated electronically over the Internet. Almost all of the information representing the content of these virtual networks is in the form of textual documents. Thus, there are literatures supporting innovation – but literatures far different in size, uniformity, tagging, and searchability than is a collection like MEDLINE.

This presents new research challenges. Key will be identifying the most appropriate kinds of search tools to uncover the unintended applications of, or modifications to, technologies. Search engines supporting two-stage retrieval from A- and then B-literatures promote a type of analogical searching in customary literature-based discovery. Additional tools are needed to support innovation discovery. There is a need to generalize from a small number of textual descriptions so that an innovation's potential value is not "obscured" by a precise description. From these broadened descriptions, tools for finding appropriate new contexts would be useful as well. One can imagine tools with certain resemblances to ARROWSMITH providing assistance in circumstances where one is seeking to understand more fully potential connections betw een an innovation and a new context. The challenge is to understand the nature of literature-based discovery in the context of linking West and South, wealthy and poor, innovation and commerce – for the betterment of all.

References

1. Amabile, T. (1988) A Model of Creativity and Innovation in Organizations. In Staw, B. and Cummings, L. (Eds.), *Research in Organizational Behavior*, vol. 10, pp. 123–167. JAI Press, Greenwich, CT
2. Arora, A. and Gambardella, A. (1994) Evaluating Technological Information and Utilizing It: Scientific Knowledge, Technological Capability, and External Linkages in Biotechnology. *Journal of Economic Behavior and Organization* 24: 91–114

3. Beyer, J. M. and Trice, H. M. (1978) *Implementing Change*. Free Press, New York
4. Bower, J. L. and Christensen, C. M. (1997) Disruptive Technologies: Catching the Wave. In John S. B. (Ed.), *Seeing Differently: Insights on Innovation*. A Harvard Business Review Book. Harvard Business School Press, Boston
5. Boyle, J. (2003) The Second Enclose Movement and the Construction of the Public Domain. *Law and Contemporary Problems* 66: 33–74
6. Callon, M. (1986) The Sociology of Actor-Network: The Case of Electric Vehicle. In Callon, M., Law, J. and Rip, A. (Eds.), *Mapping the Dynamics of Science and Technology. Sociology of Science in the Real World*, pp. 19–34. Macmillan Press, London
7. Carney, D. (1998) *Sustainable Rural Livelihoods: What Contribution Can We Make?* Department for International Development (DFID), London
8. Chambers, R., Pacey, A. and Thrupp, L. A. (Eds.) (1989) *Farmer First: Farmer Innovation and Agricultural Research*. IT Publications, London
9. Christensen, C. and Stuart, H. (2002) The Great Leap: Driving Innovation from the Base of the Pyramid. *MIT Sloan Management Review* 44(1): 51–56
10. Christensen, C., Scott, A. and Erik, R. (2004) *Seeing What's Next: Using the Theories of Innovation to Predict Industry Change*. Harvard Business School Press, Boston
11. Coleman, J. (1988) Social Capital and the Creation of Human Capital. *American Journal of Sociology* 94 (suppl): 95–120
12. Flora, J. L. (1998) Social Capital and Communities of Place. *Rural Sociology* 63(4): 481–506
13. Freeman, C. (1991) Networks of Innovators: A Synthesis of Research Issues. *Research Policy* 20: 499–514
14. Fujimura, J. (1992) Crafting Science: Standardized Packages, Boundary Objects, and 'Translations'. In Pickering, A. (Ed.), *Science as Practice and Culture*, pp. 168–211. University of Chicago Press, Chicago
15. Gordon, M., Lindsay, R. K. and Fan, W. (2002) Literature-Based Discovery on the World Wide Web. *ACM Transactions on Internet Technology* 2(4): 261–275
16. Hart, Stuart L. (2005) capitalism at the Crossroads: The Unlimited Bussiness Opportunities in Solving the World's Most Difficult Problems. Person Education, Inc. Upper Saddle River, N.J
17. Kanter, R. M. (1988) When a Thousand Flowers Bloom: Structural, Collective, and Social Conditions for Innovation in Organization. In Staw, B. M. and Cummings, L. L. (Eds.), *Research in Organizational Behavior*, vol. 10, pp. 169–211. JAI Press, Greenwich, CT
18. Keating, P. and Cambrosio, A. (2003) *Biomedical Platforms*. MIT Press, Cambridge MA
19. Krishna, A. (2002) *Active Social Capital. Tracing the Roots of Development and Democracy*. Columbia University Press, New York
20. Kuczmarski, T. (1995) *Innovation: Leadership Strategies for the Competitive Edge*. McGraw-Hill/Contemporary Books, New York, NY
21. Latour, B. (1987) *Science in Action. How to Follow Scientists and Engineers Through Society*. Open University Press, Milton Keynes
22. Lodge, G. and Wilson, C. (2006) *A Corporate Solution to Global Poverty: How Multinationals Can Help the Poor and Invigorate Their Own Legitimacy*. Princeton University Press, Princeton, NJ
23. Millennium Villages Project. http://www.earthinstitute.columbia.edu/mvp/
24. Mowery, D. C. and Rosenberg, N. (1989) *Technology and the Pursuit of Economic Growth*. Cambridge University Press, New York
25. Nord, W. R. and Tucker, S. (1987) *Implementing Routine and Radical Innovations*. Lexington Books, Lexington, MA
26. Powell, W. W. (1990) Neither Market Nor Hierarchy: Networks Forms Of Organization. In Staw, B. M. and Cummings, L. L. (Eds.), *Research in Organizational Behavior*, vol. 12, pp. 295–336. JAI Press, London
27. Powell, W. W., Koput, K. W. and Smith, K. (1996) Inter-Organizational Collaboration and the Locus on Innovation: Networks of Learning in Biotechnology. *Administrative Science Quarterly* 41: 116–145
28. Prahalad, C. K. and Hart, S. (2002) The Fortune at the Bottom of the Pyramid. *Strategy Business* 26: 54–67

29. Prahalad, C. K. (2005) The Fortune at the Bottom of the Pyramid: Eradicating Poverty through Profits. Pearson Education, Inc. Upper Saddle River, NJ
30. Putnam, R. D., Leonardi, R. and Nanetti, R. Y. (1993) *Making Democracy Work: Civic Traditions in Modern Italy*. Princeton University Press, Princeton, NJ
31. Rothwell, R. (1977) The Characteristics of Innovators and Technically Progressive Firms. *R&D Management* 7(3): 191–206
32. Schumpeter, J. A. (1942) *Capitalism, Socialism and Democracy*. Unwin, London
33. Star, L. S. and Griesemer, J. R. (1989) Institutional Ecology, 'Translations' and Boundary Objects: Amateurs and Professionals in Berkeleys Museum of Vertebrate Zoology, 1907–1939. *Social Studies of Science* 19: 387–420
34. Swanson, D. R. (1989) Online Search for Logically-Related Noninteractive Medical Literatures: A Systematic Trial-and-Error Strategy. *Journal of the American Society for Information Science* 40:356–358
35. Teece, D. J. (1992) Competition, Cooperation and Innovation: Organizational Arrangements for Regimes of Rapid Technological Progress. *Journal of Economic Behavior and Organization* 18: 1–25
36. Tornatzky, L. G. and Fleischer, M. (1990) *The Process of Technological Innovation: Reviewing The Literature*. National Science Foundation, Washington, DC
37. Uphoff, N. (Ed.) (2002) *Agroecological Innovations*. Earthscan, London
38. Verona, G., Prandelli, E., Sawhney, M. (2006) Innovation and Virtual Environments: Towards Virtual Knowledge Brokers. *Organization Studies* 27: 765–788
39. Von Hippel, E. (1976) The Dominant Role of the User in Scientific Instruments Innovation Process. *Research Policy* 5(3): 212–239
40. Von Hippel, E. (1988) *Sources of Innovation*. Oxford University Press, Oxford
41. Von Hippel, E. (2005) *Democratizing Innovation*. MIT Press, Cambridge, MA
42. Vos, R. (1991) *Drugs Looking For Diseases. Innovative Drug Research and the Development of the Beta Blockers and the Calcium Antagonists*. Kluwer, Dordrecht
43. Weeber, M., Klein, H., de Jong-van den Berg, L. T. W. and Vos, R. (2000) Text-Based Discovery in Biomedicine: The Architecture of the *DAD* System. In *Proceedings of the 2000 AMIA. Annual Fall Symposium*. Hanley & Belfus, Philadelphia, PA

The 'Open Discovery' Challenge

Jonathan D. Wren

Abstract One of the most exciting goals of literature-based discovery is the inference of new, previously undocumented relationships based upon an analysis of known relationships. Human ability to read and assimilate scientific information has long lagged the rate by which new information is produced, and the rapid accumulation of published literature has exacerbated this problem further. The idea that a computer could begin to take over part of the hypothesis formation process that has long been solely within the domain of human reason has been met with both skepticism and excitement, both of which are fully merited. Conceptually, it has already been demonstrated in several studies that a computational approach to literature analysis can lead to the generation of novel and fruitful hypotheses. The biggest barriers to progress in this field are technical in nature, dealing mostly with the shortcomings that computers have relative to humans in understanding the nature, importance and implications of relationships found in the literature. This chapter will discuss where current efforts have brought us in solving the open-discovery problem, and what barriers are limiting further progress.

1 Introduction

The amount of scientific literature is increasing exponentially,[1] along with most other databases in biomedicine, and there are far more papers published than any individual could ever hope to read. Furthermore, within this vast literature are many

J.D. Wren
Arthritis & Immunology Department, Oklahoma Medical Research Foundation,
825 N.E. 13th Street, Room W313, Oklahoma City, Oklahoma 73104-5005, USA
jdwren@gmail.com

[1] MEDLINE, for example, contained approximately 16 million records at the beginning of 2006, and is growing at a rate of approximately 4%/year, which currently equates to over 2,000 papers published per day.

P. Bruza and M. Weeber (eds.), *Literature-based Discovery*,
Springer Series in Information Science and Knowledge Management 15.

areas of research interest, more than any individual could ever hope to be aware of, leading to increasing specialization of research focus. This narrowing of relative awareness has not been a *barrier* to progress, but one could argue that it *limits* progress. In an age where data is generated faster than knowledge [2], it becomes increasingly important to be able to compile diverse sets of facts to identify high-impact hypotheses [3, 4]. The increasing emphasis on funding and conducting cross-disciplinary research and collaboration is, in part, a consequence of this expansion of information and necessary restriction of individual research focus. Intelligent tools are necessary to navigate, integrate and compile the diversity of available information to better advance all fields of scientific research.

1.1 Text-Based Knowledge Discovery

In 1986 Don Swanson illustrated that two areas of research could be functionally non-interactive, such that discoveries in one field could be relevant to studies in another, yet nonetheless remain unknown by researchers in either field because the fields have little or no overlap. Using a basic approach involving the pairing of keywords between literatures, he demonstrated that regions of overlap could be identified and novel discoveries made [5–9]. Intuitively, we recognize the value of the scientific literature in offering us insight into our own research. Who among us has not, at least once, read an article or attended a talk on a field unrelated to our own and subsequently left inspired with a new insight or direction for our own research? A broad perspective can be extremely valuable.

By enabling a computer to identify potential relationships within the scientific literature, it becomes possible to infer in an automated manner what is *not* known based upon what *is* known. Computers are, after all, perfectly suited to read large amounts of literature, catalog hundreds of thousands of names and synonyms, and simultaneously manipulate and track hundreds of relevant variables. It seems reasonable to stipulate that, for many areas of research with a significant body of associated literature, only a computer could gain the broadest possible perspective. Beyond the technical challenges associated with effective information retrieval (IR), the main challenges to the discovery of new knowledge are enabling a computer to identify *what* is of interest, *why* it is of interest and *how* the information will be conveyed to a human user. The intent of literature based discovery (LBD) is not to bypass the human researcher [10], but to provide a powerful supplement in assisting observation, analysis and inference on a large scale.

Although the LBD approach could be applied to many domains, efforts have thus far focused on the biomedical literature, specifically MEDLINE records. In part this is because MEDLINE records are freely available in electronic format, but also because most LBD efforts identify co-occurring terms as tentative relationships, whether these terms are names or medical subheadings (MeSH). Thus the nature of the association is usually non-specific and is best suited towards associations that are more general in nature. For example, when a gene is mentioned in an abstract with a disease, there is a good probability that the gene is somehow related to the disease (or suspected to be). Furthermore, if two diseases are frequently mentioned with the same genes, then it not unreasonable to assume that the diseases are related in either their pathogenesis or phenotypic characteristics. The nature of each relationship may not matter as much as the frequency of their association for such inferences. When the nature of the relationship is critical to drawing inferences, then more sophisticated methods will be necessary. For example, if mining a legal/criminal database to find names frequently associated with crimes, the nature of each association is critical to drawing any conclusions – is the person an ordinary citizen, a lawyer prosecuting cases, a judge or a policeman?

During LBD, identifying relationships that are *known* (Fig. 1(1.1)) enables one to infer relationships that are not known, yet potentially *implicit* from the relationships shared by two objects (Fig. 1(1.2)). These shared relationships provide a means to both research and justify the existence of a potentially novel relationship not explicitly contained within the literature. By comparing shared relationship sets identified within the MEDLINE relationship network against what could be expected from a random network model with the same properties, we are able to assign a statistical significance value to any given grouping of relationships (Fig. 1(1.3)).

The approach outlined in Fig. 1 is what has become known as the "open discovery" model [11, 12]. It is also sometimes referred to as "Swanson's ABC discovery model", named because the first input node (black) is referred to as the "A" node, the direct relationships (gray) are referred to as the "B" nodes and the implicit relationships (white) are referred to as the "C" nodes. These implicit relationships have also

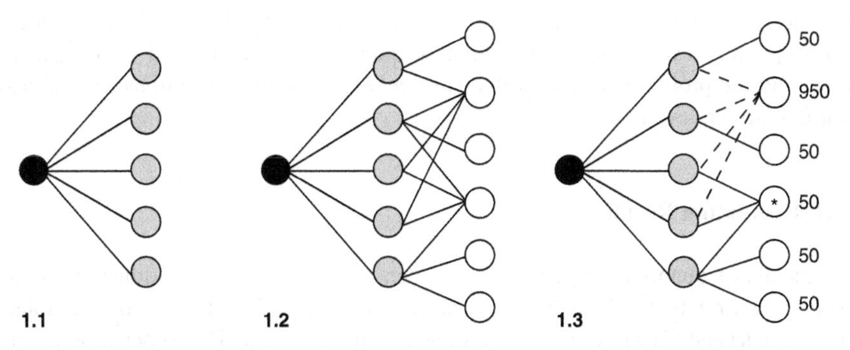

Fig. 1 Using literature-based relationships to engage in the discovery of new knowledge. (**1.1**) Beginning with an object of interest (*black node*), tentative relationships are assigned to other objects (*gray nodes*) when they are co-mentioned within MEDLINE records. (**1.2**) Each related object is then analyzed for its relationships with other objects (*white nodes*). These nodes are not directly related to the primary node, thus they are *implicitly* related. (**1.3**) These shared relationships are ranked against a random network model to establish how many would be expected by chance alone, given the connectivity of each object in the set. In this figure a hypothetical network with 1,000 nodes is analyzed. The node with the most shared relationships (four) is itself a highly connected node (connected to 95% of the network), and thus is less noteworthy from a statistical perspective than another node that shares three relationships and is connected to only 5% of the network (marked with an *asterisk*). A statistical score must be assigned in some manner to rank each of these implicit relationships for their potential significance, such as an observed to expected (Obs/Exp) ratio. Figure reproduced from [1]

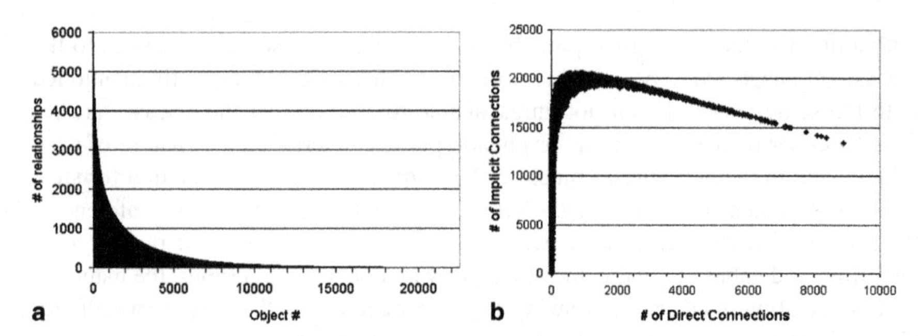

Fig. 2 Structure of the literature-based network. (**a**) The objects in a literature-based network have a disproportionate number of relationships, following a scale-free distribution. (**b**) In the case of the scientific literature, this leads to "extremely small world" network behavior by which most objects in the network are related by at least one intermediate. Figure reproduced from [1]

been referred to as "indirect" and "transitive" relationships. Similarly, the relationships themselves have also been referred to as "associations" and "connections".

Swanson outlined the open-discovery approach conceptually [6], but did not actually engage in it for most of his research because of the problems it posed. Rather, he usually began with the A and C nodes already known and focused upon exploration of the B nodes. However, because the number of relationships per object follows a scale-free distribution (Fig. 2a), the number of implicit connections found by an unbounded search increases rapidly for every direct connection. Figure 2b shows how the number of implicit connections rapidly approaches the maximum number possible (the upper asymptote) given a relatively small number of direct connections [1]. Thus, everything in the database quickly becomes related to the query object and the problem quickly shifts from *finding* implicit connections to *ranking* their potential relevance.

1.1.1 Evaluating Results

One means of quantifying performance when ranking implicit relationships is to score known relationships as if they were not known. In Fig. 1(1.3), for example, the A (black) and C (white) nodes are shown as unconnected. This is because direct relationships (the B nodes) are deliberately screened out from this set. However, if they are not screened out, they too will share relationships with the A node and can be evaluated just as any other C node in the implicit list. A previous study showed that weighting shared nodes (the B nodes) by how unlikely such a set would be shared by chance between two nodes correlated with the probability a relationship was known as well as with the strength of the relationship (Fig. 3).

This problem has been addressed by ranking implicit relationships by their connectivity within a network [1], then by attempting to extend mutual information measure (MIM) calculations from direct relationships to implicit relationships [13] and also by using fuzzy set theory (FST) to identify conceptual domains shared by

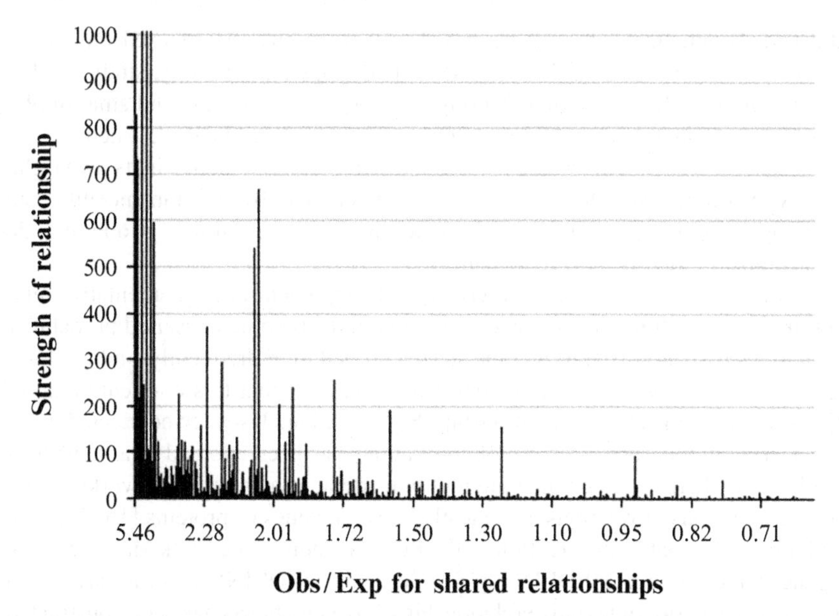

Fig. 3 The object "cardiac hypertrophy" was analyzed to identify all other objects in the database that share literature relationships with it. When a relationship is known (i.e., it has appeared in a MEDLINE title/abstract), a line is plotted on the *y*-axis, which corresponds to how many times the relationship was mentioned in MEDLINE. When the relationship is not known, there is a gap (not all gaps are visible due to *x*-axis compression). Note that frequently mentioned relationships tend to receive high scores when comparing the number of observed relationships shared by two objects to the number of relationships expected by chance (Obs/Exp). Figure reproduced from [1]

two objects [14]. Each approach had its strength and weaknesses in ranking inferences. For example, the FST approach was superior at identifying general concepts (e.g., migraines are associated with pain) whereas the MIM approach was superior at identifying more specific, informative relationships (e.g., migraines are associated with sumatriptan, a medication used to treat migraines). Regardless of the approach used, however, one major problem persisted: The amount of time the user had to spend to identify interesting implied relationships from within the set. This problem is not unique to just the studies mentioned, but rather is a general limitation of LBD in general. Relationships are defined by association and can thus be vague in their nature.

1.2 General Approach

MEDLINE abstracts contain a historical summary of biomedical discovery, and are available in electronic format free of charge from the National Library of Medicine (NLM). Abstracts are typically written without specific format or standardization of content, but are intended to convey the most pertinent aspects of the study being published. Biomedical interests are broad, yet predominantly focused on several areas of primary interest: Genetics, disease pathology and etiology, study of phenotypes, and the effects and interactions of chemical compounds and small molecules. Recognizing relevant entities or "objects" within these databases such as gene names, diseases, chemical or drug names, and so forth is a challenge in its own right. Using MeSH terms, which are assigned by curators, can bypass nomenclature and ambiguous acronym problems but MeSH terms are limited in their scope (e.g. do not encompass most specific gene names).

As objects are co-cited within a record, LBD approaches assign a tentative relationship, and sometimes a confidence score that reflects some measured probability the relationship is non-trivial. As objects are co-cited more frequently, and/or closer together (e.g. the same sentence), confidence increases that this co-mentioning of objects reflects a meaningful relationship (Fig. 4). All analyses are conducted using this uncertainty measure. This use of co-citations has been adopted in a number of experiments where an automated attempt is made at constructing networks of potential interactions or relationships, mostly between genes or proteins [15–20]. The best known is probably the creation of the PubGene genetic network via co-citation of gene names within MEDLINE [15]. Once all MEDLINE records have been processed, a network of tentative relationships between objects has been constructed and can be analyzed. The method has been applied to MEDLINE, but is extensible to any other domain where discussion is constrained to a focused summary (e.g. an

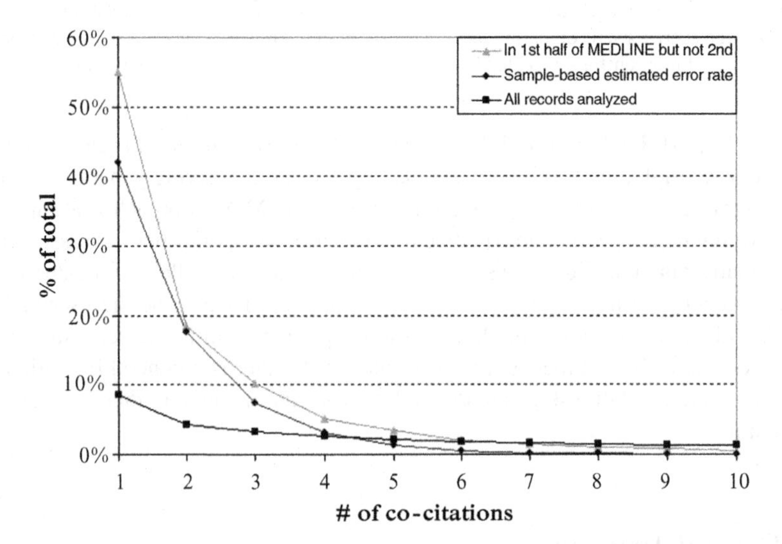

Fig. 4 Analysis of the uncertainty function in assigning tentative "relationships" based upon co-citation. *Top line* represents co-cited objects found within the first half of the 12 million MEDLINE records, but not the second half. Immediately below is the probability the uncertainty function (derived from sample-based error rates) assigns to co-cited relationships based upon the number of co-citations observed. For comparison, the overall distribution in the number of co-citations is shown at *bottom*. Figure reproduced from [1]

abstract) and co-occurrence of terms correlates with the presence or potential presence of a relationship between them (e.g. companies and products, legal precedents and key phrases such as ‘workers compensation’, etc.).

1.3 Previous LBD Applications

Open-discovery approaches have been applied to several different research problems, for example to identify compounds implicitly associated with cardiac hypertrophy, a clinically important disease that can develop in response to stress and high blood pressure. By examining the relationships shared by cardiac hypertrophy and one of the highest scoring implicitly associated compounds, chlorpromazine, it was anticipated that chlorpromazine should reduce the development of cardiac hypertrophy. It was tested using a rodent model, by giving mice isoproterenol to induce cardiac hypertrophy, with one group receiving saline injections and the other receiving chlorpromazine. Preliminary experiments suggested that chlorpromazine could significantly reduce the amount of cardiac hypertrophy induced by isoproterenol [1].

1.3.1 Type 2 Diabetes

Another analysis example involved Type 2 Diabetes, also known as Non-Insulin Dependant Diabetes Mellitus (NIDDM), and revealed a line of literature relationships that suggest the pathogenesis of NIDDM is epigenetic (Fig. 5). The analysis furthermore revealed the likely tissue of pathogenic origin (adipocytes), and narrowed the set of potentially causal factors to a general class of compounds (pro-inflammatory cytokines) implicated in the phenotype. Currently, the epigenetic

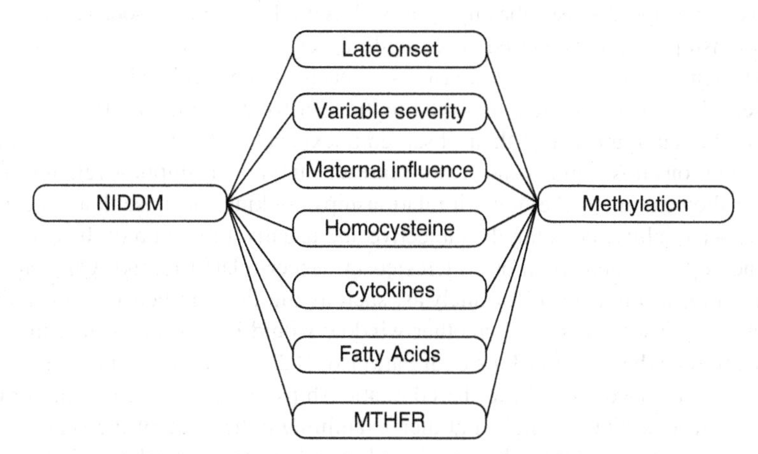

Fig. 5 A program called IRIDESCENT identified critical relationships shared by loss of DNA methylation and NIDDM (not all relationships shown), suggesting a relationship between the two

hypothesis remains untested, but seems to be gaining traction as mutation-based (e.g., single nucleotide polymorphism) models and the "complex disease" hypothesis have difficulty explaining certain observations about the etiology of NIDDM (e.g., why it is on the rise faster than population growth).

Based upon past developments and current research, it seems reasonable to presume that the ultimate goal of LBD research is the development of an intelligent system able to assimilate information in an automated manner, analyze facts and relations therein, and return to the user a set of logical conclusions and suggested courses of action based upon the current state of knowledge.

1.4 Improving on Co-Occurrences

Eventually, to provide a more targeted means of analysis, it will be necessary to expand the open-ended knowledge discovery model to include the nature of relationships in some manner. The general associative model is unfortunately too cumbersome to use, and it is difficult to rigorously test because it makes no predictions as to the nature of relationships. Thus, it is possible that every predicted implicit relationship would be true if one adopted a very lenient definition of the term "relationships". Natural language processing (NLP) provides a means of pinpointing the possible nature of the relationship between co-occurring terms (e.g. A upregulates B, B binds C). Thus it is possible that NLP could be used for the prediction of complementary and antagonistic relationships between unrelated terms.

The current LBD approaches can be summarized as general associative ones – "guilt by association" approaches. Despite their initial successes, there is still room for improvement. Figure 6 shows a general overview of the process assisted by IRIDESCENT as a generic example of how a user would explore potentially novel relationships identified by LBD approaches. First, the user selects an object for analysis. Here, the disease fibromyalgia is chosen. The literature-derived network of relationships is then queried to compile a set of terms related to fibromyalgia and then another set of relationships to each of these related terms (the implicit set). The terms are then displayed to the user for examination. Here, they are sorted in descending order of their observed to expected ratio. Gray rows represent known relationships while white rows represent unknown, implicit relationships. The user then examines the implicit relationships, looking for those that appear interesting – a quality that is highly subjective and usually a function of the examiner (e.g. oncologists would be more interested in cancer-related terms). Once an implicit relationship is chosen for analysis, such as the first implicit relationship on this list, "Parkinson's disease", another window would be opened so that the user could examine what relationships both Parkinson's disease and fibromyalgia share. The user can then examine these shared relationships, once again searching for one or more that look "interesting", and then examining either side of the shared relationship. Here, for example, the user could examine the A–B relationship, which in the window shown would be the relationship between fibromyalgia and females.

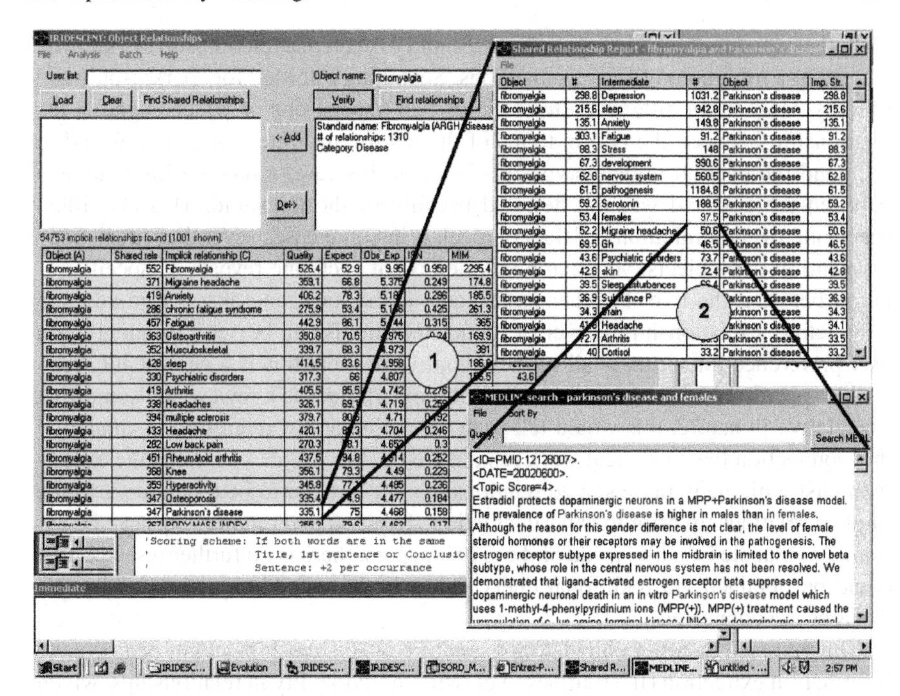

Fig. 6 Using an open-discovery approach to identify implicit relationships and explore shared relationships identified within the literature

The literature associated with this relationship is not shown here, but the nature of this relationship is that approximately 90% of fibromyalgia sufferers are female. Then, examining the corresponding B–C relationships, between females and Parkinson's, shown here in window #2 would pull up something like the next inset window. Examining the literature, with keywords highlighted for convenience, it is apparent that males disproportionately suffer from Parkinson's. Thus, at this point, the user understands one of the aspects of the implied relationship between fibromyalgia and Parkinson's. Users would then examine each of these shared relationships, one by one, to get a better idea of the overall nature of the implied relationship. This last step is the hardest since each individual relationship (e.g., of a disease to gender) may or may not paint a cohesive picture for any overall implied relationship between the two terms. It is entirely possible, if not likely, that many of the bridging B terms may be simple, isolated relationships that do not contribute at all towards an overall relationship between A and C. Thus, it could be confusing for users to try to iteratively construct a picture of a general relationship piece by piece since some of those pieces may only make sense after further analysis while others may not contribute at all towards a general A–C relationship. Subjective interpretation and a limited understanding of the nature of implied relationships are the biggest current barriers to LBD.

In the example shown in Fig. 6, the threshold to declare a relationship as "known" was set to a minimum of four co-mentions. Searching PubMed for "parkinson's and fibromyalgia" in the title or abstract yields two papers, one of which suggests the relationship between the two in terms of the neurotransmitters that are affected in each and the overlap in phenotypes [21]. So, in this case, a relationship is known between the two and was not detected because of the threshold. This also illustrates one of the limitations of the approach – in some cases several abstracts may co-mention two objects, yet examining the text of each one reveals no specific relation between the two. In other cases such as this one, one abstract co-mention may define a relationship. Lower or higher thresholds can be set depending upon user preference for virtually all of these approaches, but this is a persistent caveat. In this specific case, because the co-mentioning article was a review and somewhat speculative in nature, this would tell to the experimentalist interested in validating this connection that empirical work remains to be done. It also provides the experimentalist with many more shared relationships for him/her to better understand the implied relationship prior to experimentation. These shared relationships can be extremely valuable because, aside of these two papers, there is no further research that could be obtained via traditional query methods that would explain how the two diseases are connected.

The first step in better elucidating the nature of relationships might be to enable information extraction (IE) routines to classify directionality in relationships, which could lead to inference of complementary and antagonistic relationships. Figure 7 examines a hypothetical implicit relationship identified by an IE-based open discovery approach, with Fig. 7a showing the current approaches: Commonalities (B_1 through B_6) are identified between two objects (A and C). It is not known what type of relationship is implied by these common relationships until the user examines the text the relationships were identified in (as shown in Fig. 6). This examination can take a significant amount of time. For example, when a tentative relationship between Type 2 Diabetes and Methylation [22] in a previous analysis, although the initial implication was suggested relatively quickly, it took about 2 weeks worth of exploring the connecting relationships to better understand and identify the key components of the implied relationship. Much of this analysis is weaving a growing set of facts into a cohesive summary of what they mean collectively, which includes a willingness to look for both positive and negative evidence as well as judge what weight should be assigned to any observations that appear contradictory given all the other compiled observations. This would not be as much of a problem if it weren't for the fact that many implicit relationships are often examined before one of potential interest is found. A means of summarizing the nature of each implied relationship would be of great assistance.

Where possible, an IE-based approach to LBD would extract the nature of the relationship between objects (e.g., A affects B, but not the other way around). This directionality combined with regulatory information provides a means of inferring the general nature of relationships prior to their examination. For example, in Fig. 7c, we see that A positively affects the intermediates B_1, B_2, and B_5. In turn, these same objects positively affect C. The other intermediate relationships do not immediately

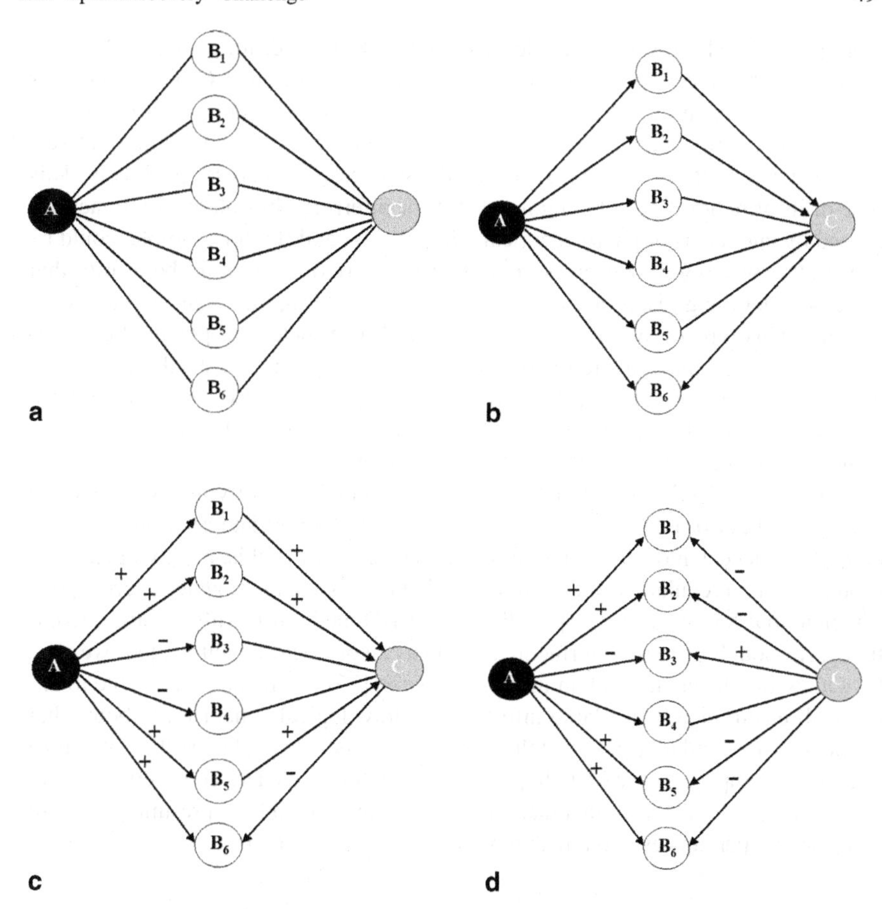

Fig. 7 Relationships identified within text and how IE would change the nature of analysis. (**a**) Current methods of ranking potentially interesting discoveries (i.e., undocumented relationships) rely upon statistical methods that suggest more relationships are shared than would be expected by chance. Here, it is unclear what the nature of the proposed relationship between objects A and C is until a user examines all the A–B and B–C intermediates. (**b**) By incorporating directional information (e.g., A affects B), greater information content is provided to the researcher. Here, for example, A appears to be affecting C through intermediates. (**c**) When information is extracted regarding the nature of relationships (e.g., A increases B), this enables inferences to be made regarding complementary and antagonistic relationships (e.g. A should increase C). (**d**) Multiple types of inferences can be made with this new model, here neither A nor C is predicted to affect the other, but rather they are anticipated to have opposing effects upon their intermediates. Notice that not all relationships necessarily have directionality or information concerning effects

provide information on how A affects C through them, if at all, but neither do they provide any contradictory information. Using this information, and without having to examine the underlying relationships beforehand, we can infer that A positively affects C.

Such a system could potentially be quite an improvement over previous methods, provided certain issues could be resolved. It would provide several ways that more generalized information could be obtained. In Fig. 7d, for example, A and C are related, but the implied relationship is not between A and C but rather their intermediates. A and C apparently have antagonistic relationships with five out of six of their intermediates. A affects B_1, B_2, B_5 and B_6 positively while C affects them negatively. It also affects B_3 negatively whereas C affects B_3 positively. This type of information would be highly useful for inferring physiological interactions caused by chemicals or pharmaceuticals. B_1, B_2, B_5 and B_5, for example, could be heart rate, sweating, blood pressure and vasoconstriction. A could be a drug that increases them (e.g., isoproterenol) and C could be a drug that reduces them (e.g., valium). This type of system could be very useful for detecting potential drug interactions. If the antagonistic relationships here were positive instead (e.g., C was ephedrine instead of valium), then this would suggest these two drugs should not be given together. Except in a case where neither one alone had sufficient effect or some enhanced effect was deliberately being sought.

Using IE to identify the nature of relationships entails identifying regulatory and associative keywords within text and assigning the appropriate relationship. Several efforts have demonstrated the feasibility and efficacy of this, mostly in terms of protein–protein regulatory interactions [23, 24] but also in more generic terms [16]. The potential for advances in open discovery LBD methods is truly exciting. Eventually, if enough of these intermediate analysis steps could be automated, we may be witnessing the creation of an in silico scientist [25] – software that is able to analyze all electronically available information, draw logical conclusions about what is both possible and plausible and then propose the most logical and efficient course of action to empirically validate hypothesized relationships derived purely in silico. Of course, we are far from that day, but it is not unreasonable to presume it is both possible and perhaps even realizable within a generation or so.

1.5 Using History as a Guide to the Future

The historical discovery of new relationships within MEDLINE abstracts and provides a benchmark dataset for knowledge discovery. It could be argued that any individual experimental validations of relationships predicted by any knowledge discovery method are somewhat anecdotal. That is, a significant amount of user-based decision goes into ascertaining what novel relationships are worth pursuing. Currently, it is not at all clear which LBD approaches are most efficient due to a lack of quantitative methods and gold standard test sets for analysis. One possible way of addressing this might be to turn to a historical analysis. If historical relationship networks could be created, we could study how they have evolved over time, asking the critical question: How many scientific discoveries known today would have been highly ranked inferences in the past – based solely upon what was known at the time? More specifically it can be asked how well any particular approach would have performed historically in predicting the probability an implicit relationship will be of future scientific relevance.

In general, scientific discovery falls roughly into one of two categories: Fortuitous and logic-based. *Fortuitous discoveries* are those that arise unexpectedly or by accident. Some might argue that, given the benefit of retrospective hindsight, some fortuitous discoveries might have been anticipated. However, the way this term will be used here is to denote discoveries that could not have been reasonably anticipated given the state of knowledge at the time of discovery. Viagra (sildenafil) is an example of a fortuitous discovery, having been originally developed as a potential treatment for angina, but instead had blockbuster success in treating erectile dysfunction (ED) [26]. The new application for ED was originally observed as a side effect during clinical trials, and while it may now make sense in terms of what is now known about sildenafil's physiological/molecular actions, it is not amenable to computational analysis because the alternative use was published before the original, intended use [26]. Rogaine (minoxidil) shares a similar history with Viagra in that it was originally developed to combat high blood pressure [27], but was discovered later to be a successful treatment for baldness [28]. Between its initial reporting in the literature in 1973 and its later use discovered around 1980, studies were published concerning its pharmacological/molecular actions that might have suggested an alternative use was possible.

Logic-based discoveries occur when an expert postulates that a new relationship can be identified (or ruled out) based upon what is currently known. Whether the expert anticipates the exact answer or not, there is a rationale for both choosing and designing the experiment such that more information can be obtained about the system in question. Conceptually, this is what most knowledge discovery approaches attempt to do: To better understand an area of research (Fig. 1(1.3), black node), unknown variables (Fig. 1(1.3), white nodes) are studied in the context of known variables (Fig. 1(1.3), gray nodes). Logic-based discoveries are those that are thought out and justified, at least to an extent, prior to the commitment of time and resources to further investigation. Preliminary results for a proposed research project typically confer a competitive advantage upon it because they imply a greater chance of success. In the absence of such results, researchers typically justify the proposed commitment of resources by extensive citing of research results obtained from others. In either case, future research is predicated upon current understanding.

It is reasonable to postulate that this latter type of scientific discovery, logic-based, is amenable to computational analysis and that there are numerous relationships published in the literature, shared by two unrelated objects, which suggest the existence of a relationship long before one is recognized. If it can be demonstrated that large-scale computational analysis of scientific information can identify important discoveries prior to their experimental validation, this has very important implications for scientific research in general. It would suggest that, to an extent, human awareness of relationships is a limiting factor in discovery and computational assistance would be of broad benefit to the scientific community.

Due to their relative simplicity and lack of reliance upon proprietary and computationally expensive NLP software, construction of co-citation networks have become an increasingly common way of ascertaining relationships among different

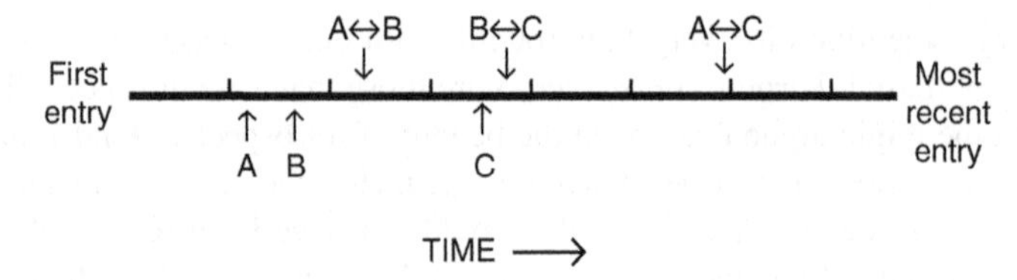

Fig. 8 Entry of objects (below timeline) and relationships (above timeline) into MEDLINE

types of objects within literature-based sources [15–20]. By this method, a relationship is "discovered" when two objects have been co-mentioned in the same abstract. However, this alone does not mean that a relationship has truly been elucidated, proposed, or understood. One could imagine a historical perspective analyzing a co-citation network of random words – one would certainly be able to identify co-citation patterns such as "red↔bird" and "bird↔house" that would predict the eventual "discovery" of the relationship "red↔house", but we can easily recognize that the nature of this relationship and prediction are trivial. This example helps in illustrating the fundamental problem in using co-citation as a metric for identifying a relationship, even when the co-citation has occurred many times: Related objects are almost unavoidably co-mentioned together, but co-mentions do not necessarily reflect a meaningful relationship.

Figure 8 illustrates graphically the variables being analyzed, with MEDLINE depicted as a time-dependant progression of published papers from the first entry to the most recent. At given points in time, the primary object of analysis, A, will first appear within the literature as will other objects such as C which will eventually be discovered to have a relationship with A. A number of intermediate factors such as B (only one is shown here for simplicity) will be related to both A and C prior to the publication of their relationship. Essentially, literature-based discovery methods are predicated upon the assumption that cases such as this exist – that at least a subset of all discoveries could have been predicted prior to their publication.

1.6 Literature Limitations

Electronically available MEDLINE records lack full experimental detail – much sequence information is not published directly in the primary literature but rather deposited into databases. Therefore knowledge discovery methods lack the ability to draw correlations between literature relationships and information contained in genomic and transcriptional (microarray) databases. Integrating experimental data with literature associations should be able to provide experimental insight in several areas.

In Fig. 9, for example, three genes within a genomic region are found to have literature correlations with a disease or phenotype. Once this is known, nearby genomic features such as CpG islands (gray square) or highly repetitive regions

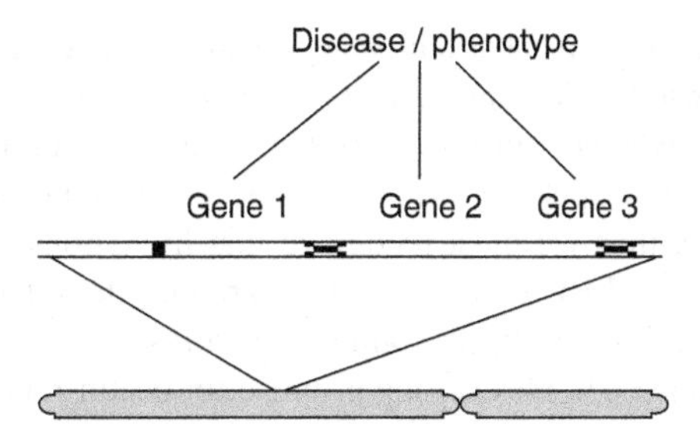

Fig. 9 Correlating literature commonalities with genomic data

(gray X) might offer a hypothesis about how recombination or silencing might contribute towards the etiology of this disease. A similar approach was conducted to identify candidate genes for diseases [29] by association of MeSH phenotypic terms with Gene Ontology (GO) terms through MeSH D terms.

1.7 Integrating Gene Expression Measurements

Within microarray experiments there are groups of genes that respond transcriptionally to changes in experimental conditions. Space limitations prevent more than a few of these genes from being mentioned within MEDLINE abstracts, so this is information that would not be obtainable the way most LBD approaches are currently implemented. However, many microarray datasets are cataloged in NCBI's Gene Expression Omnibus (GEO) [30]. A number of methods are available to cluster transcriptional responders into groups, which could then be cataloged and integrated into the literature-based network. This confers the additional advantage that implicit analyses might point directly to experimental results.

Perhaps the most difficult part of microarray analysis is not so much the cleaning, normalization and clustering of data, but ascertaining the biological relevance of the response. To do this, the researcher must first identify what is already known about the response observed within the experiment to gain confidence that aspects of their experiment correspond with previous observations. Second, and perhaps most important, they must ascertain what their experiment has told them that is not already known. The purpose of integrating microarray response datasets is to be able to answer both these questions.

2 Summary

Natural human limitations of time, expertise, speed of understanding and personal interests prevent researchers from being aware of more than a fraction of the cumulative scientific knowledge gained to date. Computers cannot yet substitute for human understanding, but can act as a mental "prosthesis" in examining and analyzing this body of knowledge. Vast amounts of time and resources have already been spent to gain this knowledge, but it has not yet been exploited for all the value it holds. Observation and perspective have always been key components in advancing science and medicine, thus we must recognize that limitations in these areas also limit the rate of progress. LBD research will help reduce these barriers and provide a broader perspective. The ability to examine networks of biomedical interactions and infer novel hypotheses holds exciting promise for health-related research.

References

1. Wren, J.D., et al., Knowledge discovery by automated identification and ranking of implicit relationships. Bioinformatics, 2004. 20(3): 389–398
2. Valencia, A., Search and retrieve. Large-scale data generation is becoming increasingly important in biological research. But how good are the tools to make sense of the data? EMBO Rep, 2002. 3(5): 396–400
3. Blagosklonny, M.V. and A.B. Pardee, Conceptual biology: unearthing the gems. Nature, 2002. 416(6879): 373
4. Bray, D., Reasoning for results. Nature, 2001. 412(6850): 863
5. Swanson, D.R., Fish oil, Raynaud's syndrome, and undiscovered public knowledge. Perspect Biol Med, 1986. 30(1): 7–18
6. Swanson, D.R., Undiscovered public knowledge. Libr Q, 1986. 56: 103–118
7. Swanson, D.R., Migraine and magnesium: eleven neglected connections. Perspect Biol Med, 1988. 31(4): 526–557
8. Swanson, D.R., Somatomedin C and arginine: implicit connections between mutually isolated literatures. Perspect Biol Med, 1990. 33(2): 157–186
9. Swanson, D.R. and N.R. Smalheiser, An interactive system for finding complementary literatures: a stimulus to scientific discovery. Artif Intell, 1997. 91: 183–203
10. Smalheiser, N.R., Informatics and hypothesis-driven research. EMBO Rep, 2002. 3(8): 702
11. Pratt, W. and M. Yetisgen-Yildiz. LitLinker: capturing connections across the biomedical literature. In Proceedings of the International Conference on Knowledge Capture (K-Cap'03), 2003, Florida
12. Srinivasan, P., Text mining: generating hypotheses from MEDLINE. J Am Soc Inf Sci Technol, 2004. 55(5): 396–413
13. Wren, J.D., Extending the mutual information measure to rank inferred literature relationships. BMC Bioinformatics, 2004. 5(1): 145
14. Wren, J.D., Using fuzzy set theory and scale-free network properties to relate MEDLINE terms. Soft Computing, 2006. 10(4): 374–381
15. Jenssen, T.K., et al., A literature network of human genes for high-throughput analysis of gene expression. Nat Genet, 2001. 28(1): 21–28
16. Rindflesch, T.C., et al., EDGAR: extraction of drugs, genes and relations from the biomedical literature. Pac Symp Biocomput, 2000. 517–528
17. Stapley, B.J. and G. Benoit, Biobibliometrics: information retrieval and visualization from co-occurrences of gene names in Medline abstracts. Pac Symp Biocomput, 2000. 529–540
18. Xenarios, I., et al., DIP, the Database of Interacting Proteins: a research tool for studying cellular networks of protein interactions. Nucleic Acids Res, 2002. 30(1): 303–305
19. Andrade, M.A. and P. Bork, Automated extraction of information in molecular biology. FEBS Lett, 2000. 476(1–2): 12–17
20. Blaschke, C., et al., Automatic extraction of biological information from scientific text: protein–protein interactions. In Proceedings of the Seventh International Conference on Intelligent Systems for Molecular Biology, 1999, pp. 60–67

21. Burgunder, J.M., Pathophysiology of akinetic movement disorders: a paradigm for studies in fibromyalgia? Z Rheumatol, 1998. 57(Suppl 2): 27–30
22. Wren, J.D. and H.R. Garner, Data-mining analysis suggests an epigenetic pathogenesis for Type II Diabetes. J Biomed Biotechnol, 2005. 2: 104–112
23. Xenarios, I., et al., DIP, the Database of Interacting Proteins: a research tool for studying cellular networks of protein interactions. Nucleic Acids Res, 2002. 30(1): 303–305
24. Zanzoni, A., et al., MINT: a Molecular INTeraction database. FEBS Lett, 2002. 513(1): 135–140
25. Wren, J.D., The emerging in-silico scientist: how text-based bioinformatics is bridging biology and artificial intelligence. IEEE Eng Med Biol Mag, 2004. 23(2): 87–93
26. Boolell, M., et al., Sildenafil: an orally active type 5 cyclic GMP-specific phosphodiesterase inhibitor for the treatment of penile erectile dysfunction. Int J Impot Res, 1996. 8(2): 47–52
27. DuCharme, D.W., et al., Pharmacologic properties of minoxidil: a new hypotensive agent. J Pharmacol Exp Ther, 1973. 184(3): 662–670
28. Zappacosta, A.R., Reversal of baldness in patient receiving minoxidil for hypertension. N Engl J Med, 1980. 303(25): 1480–1481
29. Perez-Iratxeta, C., P. Bork, and M.A. Andrade, Association of genes to genetically inherited diseases using data mining. Nat Genet, 2002. 31(3): 316–319
30. Edgar, R., M. Domrachev, and A.E. Lash, Gene expression omnibus: NCBI gene expression and hybridization array data repository. Nucleic Acids Res, 2002. 30(1): 207–210

Where is the Discovery in Literature-Based Discovery?

R.N. Kostoff

Abstract This chapter addresses the core of literature-based discovery (LBD), namely, what is discovery and how is the generation of discovery confirmed. The chapter starts with definitions of discovery and innovation, especially in the LBD context, and then proceeds to describe radical discovery and LBD. It then describes the vetting necessary to confirm the presence of discovery. Finally, the chapter concludes with a few examples where use of more comprehensive vetting techniques would have been prudent before discovery was reported. The LBD focus is on open discovery systems (start with a problem, discover a solution, or vice versa) exclusively.

1 Discovery and Innovation Definitions

Discovery is ascertaining something previously unknown or unrecognized. More formally, discovery in science is the generation of novel, interesting, plausible, and intelligible knowledge about the objects of study [42]. It can result from uncovering previously unknown information, or synthesis of publicly available knowledge whose independent segments have never been combined, and/or invention. In turn, the discovery could derive from logical exploitation of a knowledge base, and/or from spontaneous creativity (e.g., Edisonian discoveries from trial and error) [17]. Innovation reflects the metamorphosis from present practice to some new, hopefully better practice. It can be based on existing non-implemented knowledge. It can follow discovery directly, or resuscitate dormant discovery that has languished for decades.

In the LBD context, discovery is linking two or more literature concepts that have heretofore not been linked (i.e., disjoint), in order to produce novel, interesting,

R.N. Kostoff
307 Yoakum Parkway, Alexandria, VA 22304, USA
rkostoff@mitre.org

P. Bruza and M. Weeber (eds.), *Literature-based Discovery*,
Springer Series in Information Science and Knowledge Management 15.

plausible, and intelligible knowledge. Thus, simply linking two or more disparate concepts is a necessary, but not sufficient, condition for LBD. In particular, concepts may be disjoint because the value of their integration has not been recognized previously, or they may be disjoint because there appears to be little value in linking them formally. Examples of the latter (which had been proposed as potential discovery) will be shown later in this chapter.

Also, in the LBD context, innovation is the exploitation of a discovery linkage, mainly the identification of a linkage that was not being exploited at a sufficient pace.

More generally, *radical discovery* and *radical innovation* depend on the source of the inspiration and/or the magnitude of the impact. The more disparate the source of ideas from the target problem discipline, the more radical the potential discovery or innovation. The greater the magnitude of change/impact resulting from the discovery or innovation, the more radical the potential discovery or innovation.

2 Radical Discovery

Discovery and innovation are the cornerstones of frontier research. One of the methods for generating radical discovery and innovation in a target discipline is to use principles and insights from disciplines very disparate to the target discipline, to solve problems in the target discipline.

The challenge has become more critical due to increasing specialization and effective isolation of technical/medical researchers and developers [16]. As research funding and numbers of researchers have increased substantially over the past few decades, the technical literature has increased substantially as a result. Researchers/developers struggle to keep pace with their own disciplines, much less to develop awareness of other disciplines. Thus, we have the paradox that the expansion of research has led to the balkanization of research! The resulting balkanization serves as a barrier to cross-discipline knowledge transfers, and retards the progress of discovery and innovation [16].

As a result, identifying these linkages between the disparate and target disciplines, and making the subsequent extrapolations has tended to be a very serendipitous process. Until now, there has been no fully systematic approach to bridging these unconnected target and disparate disciplines.

Once the principles and associated techniques have been established for producing insights from these disparate literatures, many applications are possible. These include:

1. Promising opportunities for researchers to pursue
2. Promising new Science & Technology (S&T) directions for program managers to pursue
3. Promising leads for intelligence analysts to pursue

3 Literature-Based Discovery

The pioneering LBD study was reported in Swanson's paper hypothesizing treatments for Raynaud's Disease [36]. Many subsequent open and closed system LBD studies were performed by Swanson/Smalheiser, including migraine and magnesium [37], somatomedin-C and arginine [38], and potential biowarfare agents [40]. They also developed more formalized analytical techniques for hypothesizing radical discovery [29, 39]. Other researchers have used variants of Swanson's LBD approach for hypothesizing radical discovery in open and closed discovery systems, but only open discovery systems will be addressed here.

Gordon and Lindsay [10] used an information technology-based approach to help automate the LBD process. Weeber et al. [45] used a two step model of discovery (open discovery step followed by closed discovery step) to simulate Swanson's actual discovery. Further, Weeber et al. [46] identified potentially new target diseases for the drug thalidomide. Stegmann and Grohmann [33] used a co-word clustering of MeSH terms to identify potential discovery by location on density-centrality maps. Srinivasan [30] generated a potential discovery-identifying algorithm that operated by building MeSH-based profiles from Medline for topics. Yetisgen-Yildiz and Pratt [49] use an LBD system called LitLinker that incorporated knowledge-based methodologies with a statistical method. Van der Eijk et al. [43] mapped from a co-occurrence graph to an Associative Concept Space (ACS), to identify discovery from concepts that were close to each other in ACS but had no direct connections. Gordon and Dumais [9] used latent semantic indexing, based on higher order co-occurrences, to compute document and term similarity. Bruza et al. [5,6] generated a semantic space approach based on the Hyperspace Analogue to Language to produce representations of words in a high dimensional space. Wren et al. [48] defined classes of objects, extracted class members from a variety of source databases, and then studied their co-occurrences in Medline records to generate implicit relationships. Hristovski et al. [12, 13] used semantic predications to enhance co-occurrence-based LBD systems.

The general theory behind this approach, applied to two separate literatures, is based upon the following considerations [36].

Assume that two literatures with disjoint components can be generated, the first literature AB having a central theme "a" and sub-themes "b," and the second literature BC having a central theme(s) "b" and sub-themes "c." From these combinations, linkages can be generated through the "b" themes that connect both literatures (e.g., AB $\rightarrow$ BC). Those linkages that connect the disjoint components of the two literatures (e.g., the components of AB and BC whose intersection is zero) are candidates for discovery, since the disjoint themes "c" identified in literature BC could not have been obtained from reading literature AB alone.

For example, as shown in Swanson's initial LBD paper, dietary eicosapentaenoic acid (theme "a" from literature AB) can decrease blood viscosity (theme "b" from both literatures AB and literatures BC) and alleviate symptoms of Raynaud's disease (theme "c" from literature BC). There was no mention of eicosapentaenoic acid in the Raynaud's disease literature, but the acid was linked to the disease through the blood viscosity themes in both literatures [36].

A central problem with all the LBD studies that have been reported in the open literature is the absence of a gold standard that can be used as a basis of comparison. A true gold standard would allow comparisons of quality and quantity of potential discoveries. Many of the studies use Swanson's results (Fish Oil and Eicosapentanoic Acid) as a comparison standard. As I point out later, I have questions as to whether Swanson's hypotheses are true discoveries or are really innovations, and in any case his results give no indication of the extent of discoveries possible.

In science, if we want to estimate the quality of a predictive tool, we have a couple of main choices. If we have an exact solution to the problem, we can compare the predictive tool solution to the exact solution, and estimate the error as the difference between the exact solution and the predictive tool solution. Alternatively, if we have some way of estimating the error that accompanies a predictive tool solution, we can estimate the accuracy by that approach.

In LBD, we dont know the extent of discovery possible for any problem, and therefore are not able to estimate the comprehensiveness of any approach (recall). Further, we are not able to estimate the quality of any discovery until much testing has been done, and therefore cannot estimate the fraction of the potential discoveries identified that are in fact potential discoveries (precision).

For the LBD approaches reported in the literature, there appears to be an imbalance between the prediction of potential discovery and its validation. Most of the effort seems to have focused on the front end of the process (discovery candidate identification) with little effort on the back end (vetting of potential discovery predictions). As I will show, this has allowed non-discovery items to be represented as discovery.

As a result, I believe this insufficient vetting has contributed to the slowing of LBD implementation. LBD intrinsically has powerful capabilities, and one would have expected that, two decades after Swanson's initial paper, there would be treatments proposed for all the major chronic degenerative diseases, similar implementations for their non-medical equivalents, as well as major sponsored research programs on LBD throughout the world. As far as I know, no major clinical trials have been reported on LBD-driven hypotheses, and benefits resulting from these LBD studies have yet to be realized.

Given:

- The length of time since Swanson's pioneering paper (two DECADES)
- The massive number of medical and technical problems in need of radical discovery
- The relatively few articles published in the literature using existing LBD approaches to generate radical discovery (especially articles not published by the Swanson/Smalheiser team and not replicating the initial Raynaud's results)
- Concerns about the validity of the discoveries reported

It is clear that improvements in the fundamental LBD approach and its dissemination and acceptability are required.

My text mining group has been working on improving LBD for the past few years. The general approach we have followed was reported in 2006 [18]. We have used our specific versions of LBD on five problems (four medical, one non-medical), and have generated voluminous potential discovery for each problem. I believe we have 'cracked the code' on LBD. Our results constituted the Special Issue of the journal Technological Forecasting and Social Change, February 2008. The remainder of this chapter is focused on the potential discovery vetting procedures we have used, and includes some examples of applying our vetting procedures to discoveries that have been reported in the LBD literature.

4 Validating Potential Discovery (Vetting)

The purpose of our vetting procedures is to insure that what we report as potential discovery has not been found in the literature previously, and obeys the criteria for discovery set forth at the beginning of this chapter. If a concept has been found in the literature previously, but we believe its reporting would accelerate its development, we might report it as an innovation. We have instituted a four step vetting process that balances thoroughness with pragmatism.

The first step is to check for appearance of the potential discovery concept in the core target problem research literature. How do we define this literature? Ideally, every research document published globally in the core problem area would constitute this literature. The practical compromise we have made is to define the source literature for the core target problem literature as the Science Citation Index and Medline. While I believe this is a bare minimum core literature requirement to search for prior art/science, some examples shown in the next section illustrate that even this threshold requirement was not met before potential discovery was published.

In this first step, we operationally check for the intersection of the core target problem literature with the potential discovery literature. If the intersection is a null set, the first check is successful. Thus, if we check whether Fish Oil is a potential discovery for Raynaud's Disease, we might use the query Fish Oil (or its many specific variants) and Raynaud's Disease (or its variants), and see whether any records are retrieved. The real issue here, as will be discussed later, is how broadly or narrowly we define the core target problem literature and the potential discovery concept literature. The breadth of definition could determine whether we have generated discovery, innovation, or nothing. For example, Fish Oil may or may not be a discovery, depending on whether we define the Raynaud's Disease literature to include or exclude the Peripheral Vascular Disease literature.

The second step could be viewed as a continuation of the first step. We go beyond simple intersection to see whether there are citation linkages between the potential discovery concept and the core target problem literature that would indicate researchers were aware of the linking between these literatures previously. There are many types of citation linkages (citing papers, cited papers, papers that share common references, papers that share common citing papers, etc). Depending on how far we plan to proceed with a potential discovery (e.g., do we want to patent the potential discovery), we check at least the citing papers for linkages between the concept literature and the problem literature.

The third step is checking the patent literature. This is more difficult than the first step because of the typically wide breadth and scope of the claims in each patent.

All steps are run serially. Once the first three steps have been taken successfully, we then have the potential discovery candidate concepts examined by experts. We access two types of experts: those expert in the core target problem literature (e.g., Raynaud's Disease), and those expert in the potential discovery concept literature (e.g., Fish Oil). We ask the experts in the core target problem literature whether the potential discovery concept is indeed discovery (i.e., have they seen it before in the target problem context), and we ask the experts in the potential discovery concept literatures whether the concept could be extrapolated to the target problem. If we report potential discovery concepts that have been only partially vetted, we state that fact.

5 Examples of Validation Issues

This section presents examples of applying some of our vetting techniques to potential discoveries reported in the LBD literature.

5.1 Use of MeSH Variables

An LBD approach based on the analysis of actual text phrases is intrinsically a high-dimensional process, due to the large number of words/phrases in the literature. To circumvent this dimensionality problem, LBD researchers have used approaches that convert the problem from high-dimensional to low-dimensional. One widely used approach reported in recent LBD papers [30, 43, 49] is the use of MeSH terms instead of text terms. MeSH is a taxonomy (controlled vocabulary) in the major medical database (MEDLINE). MeSH is generated by independent indexers who read each MEDLINE article, then assign selected MeSH terms to each article. There are approximately 22,500 MeSH terms in the total MEDLINE taxonomy, orders of magnitude less than the number of text words/phrases.

The positive aspects of using MeSH terms, in addition to the reduced number of variables, are that relevant articles can be retrieved containing desired concepts but not necessarily specific text terminology. Thus, a query with a very small number of MeSH terms (e.g., lung neoplasms) can retrieve many lung cancer records that would have required perhaps hundreds of text query terms to have the same degree of coverage, and many of those retrieved records might not contain the terms lung neoplasms or lung cancer.

On the negative side, MeSH terms are restricted to the medical literature. Additionally, very recent MEDLINE records have not been indexed with MeSH terms, and would be inaccessible for LBD purposes unless text terms were added (thereby defeating one of the major reasons for selecting MeSH terms).

Further, the mapping from text terms to MeSH terms is not one-to-one, nor is it conservative like transforming from thermodynamic variables (e.g., pressure, temperature, density) to conservation variables (e.g., new variables that include combinations of the thermodynamic variables and are conserved across discontinuities, such as mass, momentum, energy) in a fluid flow system [21]. There is a well-known phenomenon called the indexer effect [11], which states essentially that indexers are fallible, and they make errors and omissions. Not all MeSH terms that should be assigned to an article are in fact assigned by the indexers. For many uses of retrievals from MEDLINE, especially where a statistical representation or a few examples are desired, the indexer effect is not overly important. However, for LBD, where any prior art/science can negate potential discovery, even one omission can prove lethal!

Thus, an algorithm that operates in MeSH space could predict discovery (where the potential discovery concept from the bc literature could not be found in the MeSH-based core ab literature), whereas the concept could be found in a text-based core ab literature. For this reason, any potential discovery made using a MeSH-based process must be vetted not only in MeSH space *but in text space as well.*

This requirement has enormous consequences! Since each MeSH term effectively represents many text terms, all these text terms have to be considered when vetting a discovery in MeSH space. Thus, *the substantive dimensional advantages that were gained in transforming from text space to MeSH space in the front end are reversed for the vetting process in the back end.* More serious is that these non-indexed or non-properly indexed records are not available for discovery using MeSH alone. To overcome this limitation, some type of text access query would be necessary.

Some examples of reported potential discoveries that were generated in MeSH space but were shown to have prior art in text space are presented in [19, 20]. To illustrate the operational mechanics of our vetting process, I will first describe in some detail one example (of many) reported in [20]. I will then summarize the single example reported in [19].

In [30], the authors generate a potential discovery-identifying algorithm that operates by building MeSH-based profiles from MEDLINE for topics. In [31, 32], the authors start with curcumin (an ingredient of the spice turmeric) and, using their algorithm, look for potential ailments this substance could benefit. Three areas identified are retinal pathologies including diabetic retinopathies, ocular inflammation and glaucoma, Crohn's Disease/Ulcerative Colitis (both members of Irritable Bowel Syndrome), and EAE/Multiple Sclerosis (MS).

I will examine the three specific claimed potential discoveries listed above using vetting steps 1 and 3, and show that the claimed discoveries are neither discovery nor innovation. Since the papers were published in 2004, and the data were taken in mid-November 2003, then potential discovery would require that no papers/patents linking curcumin and these three ailments be published prior to November 2003. My approach is to examine the core literature (papers/patents) for these three ailments published before November 2003, and ascertain whether they include curcumin as a potential treatment. If they do, then potential discovery by the authors cannot be validated.

To examine the core literature, I use text terms based on the main MeSH terms used by the author, and initially enter them (initiating topic C literature AND target A literature terms) into the PubMed search engine. This allows me to retrieve MEDLINE articles that contain the initiating topic and target literature MeSH terms and/or text terms. Then, to obtain citing or reference article data, I enter the same terms into the Science Citation Index. Finally, to obtain patent data, I enter the same terms into the Derwent Innovations Index, an aggregated global patent database on the Web of Knowledge.

Using mainly MeSH terms as text terms is a very conservative approach. If I was searching for prior art to support a legal case, I would use many other proxy terms for the initiating topic and target literatures as part of our search query. Given the breadth of coverage of the average MeSH term relative to that of the average text term, many more proxy terms could be subsumed under the average MeSH term than under the average text term. In some sense, the generality of MeSH terms relative to text terms opens the door wide for refutation of potential discovery by allowing for the implementation of large numbers of proxy terms in the vetting process.

Only a few of these examples will be shown, due to space considerations.

For the MS example, Natarajan and Bright [23] published a paper in June 2002 linking curcumin to the treatment of MS. That paper had numerous citations, five of which were published in the first half of 2003.

For the Crohn's Disease example, Sugimoto et al. [34] published a meeting Abstract in Gastroentorology in April 2002 and a research article in Gastroentorology in December 2002 [35] concluding "This finding suggests that curcumin could be a potential therapeutic agent for the treatment of patients with inflammatory bowel disease." The keywords in the research article record include Crohn's Disease and Ulcerative Colitis, and Colitis is in the title as well. See also Salh et al. [27] and Ukil et al. [41].

For the retinal pathologies example (where glaucoma focuses mainly on intraocular pressure and optic nerve damage), three examples are required due to topical diversity. For the diabetic retinopathy example, a 2002 paper [24] suggests cervistatin, pyrrolidinedithiocarbamate, or curcumin could equally serve as a treatment for proliferative diabetic retinopathy. Additionally, one of its citing papers [3] focused on the proposed curcumin treatment for diabetic retinopathy. Further, a patent whose application was published in October 2002 and which was granted in May 2003 suggested a link between curcumin and both retinopathy and Crohn's Disease/Ulcerative Colitis [2].

For the ocular inflammation example, a 2001 paper describes the use of commercially available herbal eye drops (Ophthacare) containing curcumin for a variety of infective, inflammatory and degenerative ophthalmic disorders [4]. This formulation has existed since at least the 1990s, and almost ten clinical/laboratory papers of which I am aware have been published on its evaluation between 1998 and 2002. Finally, the patent by Babish above [2] links curcumin to conjunctivitis and uveitis (an inflammation of part or all of the uvea, the middle (vascular) tunic of the eye and commonly involving the other tunics (the sclera and cornea and the retina)).

For the glaucoma example, a patent with 2001 application date and 2003 granting date links curcumin directly with glaucoma [15].

These results should not be surprising. There are over 2,300 papers in Medline related to curcumin (or curcuma or curcuminoid), of which over 20% directly address its role as an anti-inflammatory agent. Any disease in which inflammation plays a role and which is presently not co-mentioned with curcumin would be a candidate for potential discovery. Many of Srinivasan's proposed discoveries relate to inflammation-based diseases. Unfortunately, as stated previously, with many researchers working on the relation of curcumin to inflammation, the chances that the link between curcumin and a major inflammation-based disease would go unnoticed are probably small, as our vetting results seem to be showing.

What we have presented above is probably the tip of the iceberg. There are obviously other ways to refer to curcumin or Crohn's, and a search using these additional proxy terms would enhance the prior discovery. In sum, we would not call these curcumin links a discovery, or even an innovation, because the links between curcumin and retinal, intestinal, or Multiple Sclerosis problems were established well before November 2003. The algorithm under discussion, with perhaps some modifications, might be a solution for some types of semi-automating literature-based discovery, but it was not demonstrated by the three examples shown.

In [49], the authors used MeSH terms to represent document contents. They divided MEDLINE into two parts: a baseline literature including only publications before 1 January 2004, and a test literature including only publications between 1 January 2004 and 30 September 2005. They ran their algorithm LitLinker on the baseline literature and checked the generated connections in the test literature.

They reported potential discovery for three cases: Alzheimers Disease, Migraine, and Schizophrenia. They provided statistical results for all three cases, and provided one specific example of potential discovery for each of the three cases examined.

Again, I used vetting steps one and three to search the literature for references prior to 1 January 2004. For Alzheimers Disease and Migraine, I found multiple prior references, and for Schizophrenia I found a prior patent. The details are presented in [19]. In neither of the above two cases [30, 49] did I use proxy terms for either the potential discoveries or the diseases; I used only the author's own words/phrases.

Another example is the following [43]. This approach is based on mapping from a co-occurrence graph to an Associative Concept Space (ACS), where concepts are assigned a position in space such that the stronger the relationship between concepts, the closer they lie in the ACS. Potential discovery can then be obtained from strong implicit relationships, where concepts are close to each other in ACS but have no direct connections.

The authors provide two examples in [43] of ACS for small sub-sets of the total Medline database ($\ll 1\%$), whereby concepts that were close together in ACS but not connected were predicted to have a strong implicit relationship. Searching for co-occurrence of these concepts in total Medline showed a significant number of co-occurrences.

Only one of the two examples will be addressed. The authors retrieved a subset of MEDLINE records (13,423 records, February 9, 2003) from PubMed with the MeSH-based query (duchenne OR DMD OR dystrophy OR limb-girdle OR LGMD OR BMD). According to the ACS diagram, and the author's analysis, Deafness and Hearing Loss are both in close proximity to Macular Degeneration, but have no direct connections in this small sub-set of the total Medline database. Then, the authors state that a query of the whole of MEDLINE for articles containing both Deafness and Macular Degeneration yielded 28 results (June 13, 2003), some of which clearly link deafness and macular dystrophy, a condition that leads to degeneration of the macula. Thus, based on the sample results, the authors are able to predict potential discovery in the remainder of the MEDLINE database.

However, as a check, I ran the query (duchenne OR DMD OR dystrophy OR limb-girdle OR LGMD OR BMD) AND ("macular degeneration" and (deafness or hearing)) in PubMed covering text and MeSH fields, which would yield articles relating macular degeneration to hearing loss in the same subset the authors downloaded. In the sample, I found 13 pre-2003 articles that contained (macular degeneration and deafness or hearing) in the text fields and/or the MeSH fields, as opposed to the zero articles the authors claimed. All the articles linked macular degeneration/macular dystrophy to some form of hearing loss. When I re-ran the query as above minus the term 'hearing', I found 11 articles. I see no evidence of discovery, or even innovation. The known associations date back to the mid-1970s.

In all three cases [30, 43, 49], the authors would have presented much stronger arguments for their LBD approaches had they vetted in text as well as MeSH space, and presented potential discoveries that did not appear previously in the mainline literature. Or, even if prior art/science did appear as shown, they might have reported it as innovation (if it met the criteria for innovation).

5.2 Disjointness as Sufficient Condition

In the definition of discovery, the issue of disjointness of diverse literatures was addressed as follows: In the LBD context, discovery is linking two or more literature concepts that have heretofore not been linked (i.e., disjoint), in order to produce novel, interesting, plausible, and intelligible knowledge. Thus, simply linking two or more disparate concepts is a necessary, but not sufficient, condition for LBD. In particular, concepts may be disjoint because the value of their integration has not been recognized previously, or they may be disjoint because there appears to be little value in linking them formally.

Most of the LBD techniques link disparate literatures through quantity-based approaches. However, the quality of the linkages for discovery purposes requires expert judgment. The LBD community needs to be very cautious when linking a potential discovery concept from the ab concept source literature to the bc problem literature, especially in the case where there are many researchers reporting on the concept in the ab literature. What are the chances that the bc application was not perceived by at least one or two of these researchers? If the linkage were promising, why was it not reported?

I will present two examples to illustrate the problem, but they represent the tip of the iceberg for what has been reported as LBD-based discovery. In the first example, where treatments for Huntington Disease were researched, an association rules method was used to show similarities between Huntington Disease and diabetes mellitus, especially in reduced levels of insulin [12, 13]. The authors suggested (as the potential discovery) that insulin treatment might be an interesting drug for Huntington Disease. To understand the reasons for this recommendation better, I examined literatures related to Huntington Disease, diabetes, and insulin. In the Huntington Disease (HD) case, the relationship between insulin and HD should have been obvious to the HD researchers. There were some papers where HD was induced in mice, they developed diabetes, and then insulin was used to treat the diabetes. If insulin had any impact on the HD, surely the researchers would have noticed.

To validate my perceptions, I contacted an expert in Huntington Disease research, and was told that the HD problem is not an insulin deficiency problem as in type 1 diabetes, but rather an insulin release problem as in another form of diabetes. Therefore, there is no reason to expect that administering insulin would treat the HD. The key point here is that if two literatures are disjoint, there may be multiple reasons for their disjointness. It could mean that their union would produce real discovery, and no one had thought of linking them previously. Or, it could mean that their union had been considered previously, and researchers concluded that there was nothing to be gained by the linkage.

In the second example [48], the researchers searched for discovery in treating cardiac hypertrophy (defined as an increase in the size of myocites that is associated with detrimental effects on aspects of contractile and electrical function in the heart basically heart enlargement due to added physical stress on the heart muscle). Their ranking technique showed the drug chlorpromazine (CPZ) shared many implicit relations with cardiac hypertrophy, and they then inferred that it might be useful for reducing the progression of cardiac hypertrophy. There does not seem to be prior art in the journal literature, but there may be a patent that addresses the link, although it covers a wide swath.

To understand the relationship better, I examined the medical literatures on both CPZ and cardiac hypertrophy, and found the following. CPZ is a phenothiazine compound used primarily as an anti-psychotic for humans. While other phenothiazine compounds such as thioridazine have well-documented histories of strong association with cardiac arrythmias, CPZ also has a history of cardiac adverse effects on humans. Additionally, there are a large number of potential adverse side effects from the use of CPZ, including, but not limited to:

EKG changes (Particularly nonspecific Q and T wave distortions [induction of QT prolongation and torsades de pointes] – Sudden death, apparently due to cardiac arrest, has been reported); arrhythmogenic side effects caused by blockade of human ether-a-go-go-related gene (HERG) potassium channels; Neuroleptic Malignant Syndrome; neuromuscular reactions (tardive dyskinesia; dystonias, motor restlessness, pseudo-parkinsonism); convulsive seizures (petit mal and grand mal); lowered seizure thresholds; bone marrow depression; prolonged jaundice; hyperreflexia or hyporeflexia in newborn infants whose mothers received phenothiazines; drowsiness; hematological disorders, including agranulocytosis, eosinophilia, leukopenia, hemolytic anemia, aplastic anemia, thrombocytopenic purpura and pancytopenia; postural hypotension, simple tachycardia, momentary fainting and dizziness; cerebral edema; abnormality of the cerebrospinal fluid proteins; allergic reactions of a mild urticarial type or photosensitivity; exfoliative dermatitis; asthma, laryngeal edema, angioneurotic edema and anaphylactoid reactions; amenorrhea, gynecomastia, hyperglycemia, hypoglycemia and glycosuria; corneal and lenticular changes, epithelial keratopathy and pigmentary retinopathy; some respiratory failure following CNS depression; paralytic ileus; thermoregulation difficulties.

Why, then, given this history of adverse side effects, which includes some adverse cardiac side-effects, would one highlight CPZ for cardiac hypertrophy (or any cardiac problem) as a discovery to be pursued for humans? For control of psychotic problems, CPZ may be the lesser of two evils, but does that hold true for control of cardiac problems?

To validate my perceptions, I contacted two experts in cardiac hypertrophy, and was told there is no sufficient evidence that would support pursuing CPZ for treating cardiac hypertrophy in humans and link to hypertrophic cardiomyopathy was not clear.

This example illustrates the problem with using quantity-based measures to associate with quality predictions. The authors ranking method emphasizes co-occurrences and persistence of relationships. If CPZ has a persistent and frequent history of being associated with adverse cardiac effects, both directly and as a member of a class (phenothiazines) even more strongly associated with adverse cardiac effects, then it would have a strong implicit relationship with cardiac hypertrophy. The quality of the total somatic relationship is not necessarily positive, as this example shows. While the authors ran some lab experiments showing that CPZ reduced cardiac hypertrophy in mice [48], the relation may reflect a local optimization on cardiac hypertrophy, and a global sub-optimization on overall somatic well-being.

I ran a shortcut LBD analysis combining some of our methods with Arrowsmith, and found a potential discovery applicable to cardiac hypertrophy. Cereal fiber has been shown to increase circulating adiponectin concentrations in diabetic men and women [25, 26]. In turn, adiponectin, an adipocyte-derived protein, has cardioprotective actions (e.g., [Adiponectin receptors] AdipoR1 and AdipoR2 mediate the suppressive effects of full-length and globular adiponectin on ET-1-induced hypertrophy in cultured cardiomyocytes, and AMPK is involved in signal transduction through these receptors). AdipoR1 and AdipoR2 might play a role in the pathogenesis of ET-1-related cardiomyocyte hypertrophy after myocardial infarction, or adiponectin deficiency leads to progressive cardiac remodeling in pressure overloaded condition mediated via lowing AMPK signaling and impaired glucose metabolism [22]. Therefore, use of cereal fiber in the diet could potentially contribute to ameliorating cardiac hypertrophy, with probably very few or no adverse side effects, and perhaps some positive side effects.

Since our previous LBD studies have generated voluminous amounts (hundreds) of potential discovery on each disease studied, I see no reason this would be different for cardiac hypertrophy, and the single potential discovery presented here would be one of very many resulting from a full study.

5.3 Definition of Prior Art

This third validation category brings us back full circle to the definition of discovery and what is prior art. As an example, many LBD studies refer to the potential discovery of Fish Oil for Raynaud's Disease [9, 10, 12, 36, 45]. Use of Fish Oil for circulatory problems was reported in the literature at least as far back as the 1970s, and possibly even earlier. Papers in the late 1970s discussed the impact of Fish Oil on atherosclerosis [1], thrombosis [8], vascular disease [14], and papers in the early 1980s also focused on vascular disease [7] and peripheral vascular disease [47]. While none of these papers mentioned Raynaud's Disease specifically, how much of a leap is it from peripheral vascular disease to Raynaud's Disease? For example, [28] lists drug therapies for peripheral vascular disease, and presents this information in two categories: intermittent claudication and Raynaud's Disease. Additionally, most of the hospital Web sites I examined list Raynaud's Disease under peripheral vascular diseases. Thus, depending on how broadly the core Raynaud's Disease literature is defined, Fish Oil may or may not have been a potential discovery.

6 Summary and Conclusions

In summary, this chapter has shown the importance of having rigorous definitions of discovery and innovation, and using a rigorous vetting process to insure that no prior art exists. While one can always identify further sources that could be checked for prior art, nevertheless, the sources suggested in this chapter should be viewed as a threshold before reporting potential discovery in the literature, I firmly believe that one of the major roadblocks to wide-scale acceptance of LBD by the potential user community has been the lack of real discovery reported in the literature. Until more rigorous standards for defining discovery have been implemented, and more rigorous vetting techniques used, LBD will have problems in taking its rightful place in the arsenal of discovery weapons.

References

1. F. Angelico and P. Amodeo. Eicosapentaenoic acid and prevention of atherosclerosis. *Lancet*, 2(8088):531, 1978
2. J. G. Babish, T. Howell, L. Pacioretty, T. M. Howell, and L. M. Pacioretty. Composition for treating e.g. inflammation or inflammation based diseases, comprising curcuminoid species and alpha- or beta-acid. Patent Number US2003096027-A1, May 22, 2003

3. M. Balasubramanyam, A. Koteswari, R. S. Kumar, S. F. Monickaraj, J. U. Maheswari, and V. Mohan. Curcumin-induced inhibition of cellular reactive oxygen species generation: novel therapeutic implications. *Journal of Bioscience*, 28(6):715–721, 2003
4. N. R. Biswas, S. K. Gupta, G. K. Das, N. Kumar, P. K. Mongre, D. Haldar, and S. Beri. Evaluation of ophthacare eye drops – a herbal formulation in the management of various ophthalmic disorders. *Phytotherapy Research*, 15(7):618–620, 2001
5. P. Bruza, R. Cole, D. W. Song, and Z. Bari. Towards operational abduction from a cognitive perspective. *Logic Journal of the IGPL*, 14(2):161–177, 2006
6. P. Bruza, D. W. Song, and R. McArthur. Abduction in semantic space: towards a logic of discovery. *Logic Journal of the IGPL*, 12(2):97–109, 2004
7. I. J. Cartwright, A. G. Pockley, and J. H. Galloway. The effects of dietary omega-3 polyunsaturated fatty-acids on erythrocyte-membrane phospholipids, erythrocyte deformability and blood-viscosity in healthy-volunteers. *Atherosclerosis*, 55(3):267–281, 1985
8. J. Dyerberg, H. O. Bang, E. Stoffersen, S. Moncada, and J. R. Vane. Eicosapentanoic acid and prevention of thrombosis and atherosclerosis. *Lancet*, 2(8081):117–119, 1978
9. M. D. Gordon and S. Dumais. Using latent semantic indexing for literature based discovery. *Journal of the American Society for Information Science and Technology*, 49(8):674–685, 1998
10. M. D. Gordon and R. K. Lindsay. Toward discovery support systems: a replication, reexamination, and extension of Swanson's work on literature-based discovery of a connection between raynaud's and fish oil. *Journal of the American Society for Information Science and Technology*, 47(2):116–128, 1996
11. P. Healey, H. Rothman, and P. K. Hoch. An experiment in science mapping for research planning. *Research Policy*, 15(5):233–251, 1986
12. D. Hristovski, B. Peterlin, and S. Dzeroski. Literature based discovery support system and its application to disease gene identification. In *Proceedings of AMIA Fall Symposium*, p. 928. Hanley and Belfus, Philadelphia, PA, 2001
13. D. Hristovski, B. Peterlin, J. A. Mitchell, and S. M. Humphrey. Using literature-based discovery to identify disease candidate genes. *International Journal of Medical Informatics*, 74(2–4): 289–298, 2005
14. J. A. Jakubowski and N. G. Ardlie. Evidence for the mechanism by which eicosapentaenoic acid inhibits human-platelet aggregation and secretion – implications for the prevention of vascular-disease. *Thrombosis Research*, 16(1–2):205–217, 1979
15. A. Komatsu. Preparation of health drink, involves processing preset amount of dry turmeric powder, dry curcuma zedoaria powder, dry curcuma wenyujin powder and sea tangle powder with distilled white liquor at specific temperature. Patent Number JP2003189819-A, July 8, 2003
16. R. N. Kostoff. Overcoming specialization. *BioScience*, 52(10):937–941, 2002
17. R. N. Kostoff. Stimulating innovation. In L. V. Shavinina, editor, *International Handbook of Innovation*, pp. 388–400. Elsevier Social and Behavioral Sciences, Oxford, UK, 2003
18. R. N. Kostoff. Systematic acceleration of radical discovery and innovation in science and technology. *Technological Forecasting and Social Change*, 73(8):923–936, 2006
19. R. N. Kostoff. Validation of potential literature-based discovery candidates. *Journal of Biomedical Informatics*, 40(4):448–450, 2007
20. R. N. Kostoff, J. A. Block, M. B. Briggs, R. L. Rushenberg, J. A. Stump, D. Johnson, C. M. Arndt, T. J. Lyons, and J. R. Wyatt. Literature-related discovery. *ARIST*, 2008
21. P. Lax and B. Wendroff. Systems of conservation laws. *Communications on Pure and Applied Mathematics*, 13(2):217–237, 1960
22. Y. Liao, S. Takashima, N. Maeda, N. Ouchi, K. Komamura, I. Shimomura, M. Hori, Y. Matsuzawa, T. Funahashi, and M. Kitakaze. Exacerbation of heart failure in adiponectin-deficient mice due to impaired regulation of ampk and glucose metabolism. *Cardiovascular Research*, 67(4):705–713, 2005
23. C. Natarajan and J. J. Bright. Curcumin inhibits experimental allergic encephalomyelitis by blocking IL-12 signaling through janus kinase-STAT pathway in T lymphocytes. *Journal of Immunology*, 168(12):6506–6513, 2002

24. T. Okamoto, S. Yamagishi, Y. Inagaki, S. Amano, K. Koga, R. Abe, M. Takeuchi, S. Ohno, A. Yoshimura, and Z. Makita. Angiogenesis induced by advanced glycation end products and its prevention by cerivastatin. *The FASEB Journal*, 16(14):1928–1930, 2002
25. L. Qi, J. B. Meigs, S. Liu, J. E. Manson, C. Mantzoros, and F. B. Hu. Dietary fibers and glycemic load, obesity, and plasma adiponectin levels in women with type 2 diabetes. *Diabetes Care*, 29(7):1501–1505, 2006
26. L. Qi, E. Rimm, S. Liu, N. Rifai, and F. B. Hu. Dietary glycemic index, glycemic load, cereal fiber, and plasma adiponectin concentration in diabetic men. *Diabetes Care*, 28(5):1022–1028, 2005
27. B. Salh, K. Assi, V. Templeman, K. Parhar, D. Owen, A. Gomez-Munoz, and K. Jacobson. Curcumin attenuates DNB-induced murine colitis. *American Journal of Physiology-Gastrointestinal and Liver Physiology*, 285(1):G235–G243, 2003 [see also [44]]
28. SIGN. Drug therapy for peripheral vascular disease: a national clinical guideline. Technical Report IGN Publication Number 27, Scottish Intercollegiate Guidelines Network, Edinburgh, Scotland, 1998
29. N. R. Smalheiser and D. R. Swanson. Using ARROWSMITH: a computer-assisted approach to formulating and assessing scientific hypotheses. *Computer Methods and Programs in Biomedicine*, 57(3):149–153, 1998
30. Padmini Srinivasan. Text mining: generating hypotheses from MEDLINE. *Journal of the American Society for Information Science and Technology*, 55(5):396–413, 2004
31. Padmini Srinivasan and Bishara Libbus. Mining MEDLINE for implicit links between dietary substances and diseases. *Bioinformatics*, 20(suppl 1):I290–I296, 2004
32. Padmini Srinivasan, Bishara Libbus, and A. K. Sehgal. Mining MEDLINE: postulating a beneficial role for curcumin longa in retinal diseases. *HLT BioLink*, 20(suppl 1):I290–I296, 2004
33. J. Stegmann and G. Grohmann. Hypothesis generation guided by co-word clustering. *Scientometrics*, 56(1):111–135, 2003
34. K. Sugimoto, H. Hanai, T. Aoshi, K. Tozawa, M. Uchijima, T. Nagata, and Y. Koide. Curcumin ameliorates trinitrobenzene sulfuric acid (TNBS) – induced colitis in mice. *Gastroenterology*, 122(4 suppl 1):A395–A396, T993, 2002
35. K. Sugimoto, H. Hanai, T. Aoshi, K. Tozawa, M. Uchijima, T. Nagata, and Y. Koide. Curcumin prevents and ameliorates trinitrobenzene sulfonic acid-induced colitis in mice. *Gastroenterology*, 123(6):1912–1922, 2002
36. D. R. Swanson. Undiscovered public knowledge. *Library Quarterly*, 56:103–118, 1986
37. D. R. Swanson. Migraine and magnesium: eleven neglected connections. *Perspectives in Biology and Medicine*, 31(4):526–557, 1988
38. D. R. Swanson. Somatomedin-c and arginine – implicit connections between mutually isolated literatures. *Perspectives in Biology and Medicine*, 33(2):157–186, 1990
39. D. R. Swanson and N. R. Smalheiser. An interactive system for finding complementary literatures: a stimulus to scientific discovery. *Artificial Intelligence*, 91(2), 1997
40. D. R. Swanson, N. R. Smalheiser, and A. Bookstein. Information discovery from complementary literatures: categorizing viruses as potential weapons. *Journal of the American Society for Information Science and Technology*, 52(10):797–812, 2001
41. A. Ukil, S. Maity, S. Karmakar, N. Datta, J. R. Vedasiromoni, and P. K. Das. Curcumin, the major component of food flavour turmeric, reduces mucosal injury in trinitrobenzene sulphonic acid-induced colitis. *British Journal of Pharmacology*, 139(2):209–218, 2003
42. R. E. Valdes-Perez. Principles of human-computer collaboration for knowledge discovery in science. *Artificial Intelligence*, 107(2):335–346, 1999
43. C. C. van der Eijk, E. M. van Mulligen, J. A. Kors, B. Mons, and J. van den Berg. Constructing an associative concept space for literature-based discovery. *Journal of the American Society for Information Science and Technology*, 55(5):436–444, 2004
44. C. Varga, M. Cavicchi, A. Orsi, D. Lamarque, J. C. Delchier, D. Rees, and B. J. Whittle. Beneficial effect of P54, a novel curcumin preparation in TNBS-induced colitis in rats. *Gastroenterology*, 120(5 suppl 1):A691, 2001

45. M. Weeber, H. Klein, L. T. W. de Jong-van den Berg, and R. Vos. Using concepts in literature-based discovery: simulating swanson's raynaud-fish oil and migraine-magnesium discoveries. *Journal of the American Society for Information Science and Technology*, 52(7):548–557, 2001
46. M. Weeber, R. Vos, H. Klein, L. T. W. de Jong-van den Berg, A. R. Aronson, and G. Molema. Generating hypotheses by discovering implicit associations in the literature: a case report of a search for new potential therapeutic uses for thalidomide. *Journal of the American Medical Informatics Association*, 10(3):252–259, 2003
47. B. E. Woodcock, E. Smith, and W. H. Lambert. Beneficial effect of fish oil on blood-viscosity in peripheral vascular-disease. *British Medical Journal*, 288(6417):592–594, 1984
48. J. D. Wren, R. Bekeredjian, J. A. Stewart, R. V. Shohet, and H. R. Garner. Knowledge discovery by automated identification and ranking of implicit relationships. *Bioinformatics*, 20(3):389–398, 2004
49. M. Yetisgen-Yildiz and W. Pratt. Using statistical and knowledge-based approaches for literature-based discovery. *Journal of Biomedical Informatics*, 39(6):600–611, 2006

Part II
Methodology and Applications

Analyzing LBD Methods using a General Framework

A.K. Sehgal, X.Y. Qiu, and P. Srinivasan

Abstract This chapter provides a birds-eye view of the methods used for literature-based discovery (LBD). We study these methods with the help of a simple framework that emphasizes objects, links, inference methods, and additional knowledge sources. We consider methods from a domain independent perspective. Specifically, we review LBD research on postulating gene–disease connections, LBD systems designed for general purpose biomedical discovery goals, as well as LBD research applied to the web. Opportunities for new methods, gaps in our knowledge, and critical differences between methods are recognized when the "literature on LBD" is viewed through the scope of our framework. The main contributions of this chapter are in presenting open problems in LBD and outlining avenues for further research.

1 Introduction

Literature based discovery (LBD), also known as text mining and knowledge discovery from text (KDT), has garnered significant breadth and depth as a field of research and development. The field is vibrant as seen for instance by the growing number

---

A.K. Sehgal
Department of Computer Science, The University of Iowa, Iowa City, IA 52242, USA

X.Y. Qiu
Department of Management Sciences, The University of Iowa, Iowa City, IA 52242, USA

P. Srinivasan
Department of Computer Science, The University of Iowa, Iowa City, IA 52242, USA
and
Department of Management Sciences, The University of Iowa, Iowa City, IA 52242, USA
and
School of Library and Information Science, The University of Iowa, Iowa City, IA 52242, USA
psriniva@iowa.uiowa.edu

P. Bruza and M. Weeber (eds.), *Literature-based Discovery*,
Springer Series in Information Science and Knowledge Management 15.

of conferences, workshops, papers, commercial and free systems, and review papers (e.g. Weeber et al. [27]). There is also a growing variety of LBD methods, orientations and applications. We observe that in general, LBD strategies are designed to fit the problem at hand with methods selected or designed in a somewhat adhoc manner. And since the space of text mining problems is broad (and growing), the range of solutions proposed and applied is also broad. As a consequence, there is a bewildering array of methods in text mining. While this situation offers almost free rein to researchers, it also makes it challenging to determine what methods (or aspects about methods) are most successful or most appropriate for a given problem in a specific domain. Seemingly similar problems are sometimes addressed using significantly different approaches while certain LBD approaches exhibit broader appeal. At this point, what is needed is a "meta-level" examination of the major milestones in methods. Thus our goal is to begin such an examination by offering a bird's eye view of LBD research. In particular, we analyze methodologies in LBD papers using a general framework. Some of the expected outcomes from such a framework-based review are to be able to more effectively:

1. Compare and contrast research in LBD
2. Observe the gaps in research
3. Assess the prevalence of particular methods
4. Make comparisons across domains or type of text
5. Understand the relationship between LBD methods and problems being solved
6. Understand the evolution of ideas in LBD research

Although we select papers for review with a fairly broad brush, we do not claim comprehensiveness in coverage. Likely the selections will reveal our own inclinations and preferences. Despite these built-in limitations, this framework-based review, is to the best of our knowledge, a first attempt at *domain-independent* meta-analysis of LBD research with a significant emphasis on *methodology*. We offer it as a potentially useful starting point for discussion, extension and refinements by others.

2 A Framework for Analyzing LBD Research and Development

LBD refers to automatic or semi-automatic efforts supporting end user exploration of a text collection with the goal of generating or exploring new ideas. Specifically, LBD systems help *form* and/or *explore* hypotheses using large collections of texts. LBD takes off from an age old process fundamental to fields of intellectual endeavor such as the sciences, where ideas build upon prior published work. LBD systems are of interest given their potential to consider very large sets of documents as also documents from fields that a user would not normally study. Generating or exploring hypotheses within such large-scale and heterogeneous document collections typically implies effort well beyond human capacity. While offering these advantages, LBD systems are far from reflecting the human acuity involved in the manual processes they try to emulate. In fact, LBD output is always tentative, requiring end user decisions on suggestions to take forward and suggestions to reject.

The kinds of hypotheses of interest in LBD are those that somehow relate at least a pair of entities. For example, an LBD system may suggest a financial connection between two individuals, or a link between a gene and a disease, or indicate potential interest in a product from the view point of an organization, or find communities of people related in some novel way. LBD is clearly akin to data-mining from *structured* data which also focuses on hypothesis formation and knowledge discovery. The power of LBD is seen especially in its capacity to generate novel ideas by bridging different areas of specializations represented in the text collection, thus reflecting a multidisciplinary perspective.

LBD research has strong and early roots based in the research of Swanson and Smalheiser (see chapter titled 'Literature Based Discovery? The Very Idea'). Their initial LBD efforts lead them to successfully postulate several hypotheses by linking evidence extracted from different documents. However, these were accomplished through significant manual effort. Since then a growing body of research, including Swanson and Smalheiser's own work with their ARROWSMITH systems,[1,2] aims at automating LBD. The overall approach is to try to automate as many of the key steps in LBD as possible, thereby minimizing human intervention. LBD strategies have been developed and applied to biomedicine in general and bioinformatics in particular. These efforts typically involve the MEDLINE[3] database with optionally allied sources such as Entrez Gene[4] and OMIM[5] and vocabularies such as the Gene Ontology [4]. LBD has also been applied to the humanities field as well as to knowledge discovery problems on the web.

2.1 LBD Framework

Based upon our own experiences in LBD [22–24], including work on Manjal[6], our prototype biomedical LBD system, and our understanding of the literature, we propose a simple framework for analyzing LBD methods. The framework has the four dimensions listed in Fig. 1. It allows us to understand and specify the key methodological choices made by the authors of the papers reviewed. It also allows us to objectively compare studies and suggest instances where alternative methods may also be beneficial.

Objects refer to the kinds of concepts (abstract or otherwise) that are the focus of the LBD effort. In some cases these may refer to entities of a specific type such as genes, perhaps even limited to genes of a specific species. Other studies may

[1] University of Chicago version: http://kiwi.uchicago.edu/

[2] University of Illinois – Chicago version: http://arrowsmith.psych.uic.edu

[3] http://www.ncbi.nlm.nih.gov/entrez/query.fcgi?DB=pubmed

[4] http://www.ncbi.nlm.nih.gov/entrez/query.fcgi?db=gene

[5] http://www.ncbi.nlm.nih.gov/entrez/query.fcgi?db=OMIM

[6] http://sulu.info-science.uiowa.edu/Manjal.html

- Objects
- Links
- Inference methods
- Additional data/knowledge sources

Fig. 1 LBD framework

involve multiple varieties of entity types such as persons, organizations and products. In still other cases, the LBD objects may be a collection of "topics" where topics may refer to PubMed queries (e.g. ("hypertension" AND ("2001/01/01"[PDAT]: "2006/12/31"[PDAT])). Given a particular kind of object (entity), say genes, studies may differ on what information is used to derive representations for each gene. One could use, for instance, MEDLINE records related to the gene as the representation, or the gene's sequence, or its MeSH profile, or the MEDLINE sentences in which the gene name or its alias appears. Even with a given source, say MEDLINE records, variations are possible. One may retrieve records from PubMed using the disjunction of the gene's various names. Or, one may use only those documents that provide evidence of GO based annotation for the gene. Additionally, weights are sometimes allocated to the different features in the representation. One document (or a sentence or a MeSH term) may be more central to the gene than another. And of course different studies may employ different weighting strategies, including none at all. Thus while analyzing the success (or failure) of LBD methods for specific problems, one has to pay careful attention to the kinds of objects and their representations used.

Links represent associations between objects of interest. Links may vary from straightforward co-occurrence based connections to similarity-based assessments to more semantically motivated relationships. These may be obtained in different ways, e.g., from curated or automatically generated databases or extracted from texts using pattern recognition or more advanced NLP methods. As with objects, links may also be weighted. Additionally these may be directed and/or labeled. Multiple links between objects may also be used. Each of these options and their various combinations offer different capabilities to an LBD system.

Inference methods refer to the reasoning strategies used to identify *implicit* connections between objects. In the simplest case, one may use a transitive relationship between two objects to infer a novel connection. Extensions of this idea lead to the classic strategies of Open and Closed discovery (see chapter titled 'Literature Based Discovery? The Very Idea'). Other methods are also available. For example, connections may be inferred between two objects if their representations (retrieved documents, MeSH profiles) are very similar even if they do not co-occur.

The final dimension refers to whether *additional sources* are used within the LBD process. For example, sequence data or disease mapping to chromosomal regions may be used to constrain the LBD hypotheses generated regarding putative links between genes and diseases. Similarly, geographical co-location may be used to limit the potential customers of products suggested by the LBD process.

To close this section on our framework, it may be that two studies with different goals use methods that are quite similar. This would indicate the generalizability of the methods. On the other hand two methods applied to the same goal could look very different. This might imply flexibility in the problem. It may also call for or lead to direct comparisons of the two methods. More generally, the framework might also point to dimensions that are less well explored than others. Thus our goal is also to identify open areas for research on LBD.

We now analyze select LBD research using our framework. Reviews of LBD methods are done primarily based upon descriptions in published papers. We present our analysis in three parts. In Sect. 3 we analyze a set of papers that directly target the discovery of gene–disease connections. By focusing on this subset of LBD research we will highlight the variability across methods even when they have the same LBD goal. In Sect. 4 we study general purpose LBD systems in biomedicine. Finally in Sect. 5 we examine LBD applications on the web. Each part includes an analysis of methods covered. Following this, in Sect. 6 we present our conclusions.

3 LBD for Postulating Gene–Disease Connections

Postulating novel connections between genes and diseases is a major emphasis in bioinformatics text mining. In all papers reviewed in this section, gene–disease links are postulated without qualification as to the type of link. For each study reviewed (throughout the paper) we identify its major features in terms of the key dimensions of our framework. We present Objects, Links and Inference Methods in a table. Additional knowledge sources are described in the discussions. By default, links are considered weighted, symmetric and unlabeled. Otherwise, unweighted links are marked with a U in the Notes column, asymmetric weights with an A and labeled links with an L. These qualifications under Notes apply only to the links.

3.1 G2D (Perez-Iratxeta et al. 2002)

G2D [20] is a system that ranks candidate genes for genetically inherited diseases for which no underlying gene has yet been assigned. The key objects and link are shown in Table 1. Two types of links are core to their procedure. Although both involve MEDLINE as the source, records supporting the links are extracted in different ways. The first link type (L1) associates 'pathological conditions' and

Table 1 G2D – Perez-Iratxeta et al. (2002)

Type ID	Object type	Object representation/link derivation	Notes
O1	Disease	Disease manifestations (category C MeSH terms)	–
O2	Chemical	Chemical (category D MeSH terms)	–
O3	Annotation	GO term	–
O4	Gene sequence	From RefSeq	–
O5	Gene sequence	From chromosome region of disease	–
L1	O1, O2	Co-occurrence in MEDLINE records about disease of interest	
L2	O2, O3	Co-occurrence between O2 and O3 in MEDLINE records used as evidence to annotate sequences O4 with O3	
L3	O1, O3	Inferred through L1 and L2	
L4	O3, O4	O3 annotates sequence O4	L
L5	O4, O5	Homology	U
L6	O1, O4	L3 and L4	
L7	O1, O5	Inferred from L5 and L6	
IM	*Average of fuzzy scores representing best paths between disease and GO terms*		

'chemical terms'. Pathological conditions are represented by category 'C' MeSH terms while chemical terms are category 'D' MeSH terms. L1 strength is a symmetric weight and is calculated as the number of MEDLINE records with both terms divided by the number of records having either term. The second link type (L2) connects 'chemical terms' and GO terms describing protein function. L2 strength is also symmetric and is calculated as the number of records having the chemical term and also providing evidence supporting annotation by the GO term in RefSeq[7] divided by the number of records with either feature.

L3 is inferred between the pathological condition and GO term pairs. Since several chemical bridges are possible between a pair, the weight is a fuzzy score representing the best possible chemical bridge. It is symmetric and is calculated as the product of weights for the best chemical path. Given that a disease may be characterized by several pathological conditions, the L3 weight between a disease and a GO term is the highest weight calculated for any of its manifestations. They rank candidate sequences using the homology between RefSeq annotated sequences in the chromosomal region to which the disease is mapped (L5). Ranking of candidate sequences to a disease is by the average of the scores calculated for each of their GO terms and the disease.

Thus when we look closely at their methods at least five types of objects and seven link types may be identified. Notice also for example, that their approach looks for the best path connecting a GO term to a disease using disease pathological conditions and chemicals as bridges. However, the score for a candidate gene is not the best offered through its annotations, but the average. This score is then normalized as an R score to allow for standardized comparisons.

[7] http://www.ncbi.nlm.nih.gov/RefSeq/

3.2 eVOC (Tiffin et al. 2005)

The authors use the eVOC Anatomical System ontology[8] as a bridging vocabulary to select candidate disease genes [26]. (We refer to this system as the eVOC system.) Specifically they exploit information on the genes' expression profiles within tissues affected by the disease of interest. As described by them, researchers may mine associations between disease and affected tissues without having a clinical understanding of the disease. This connection may then be applied to the selection of candidate genes for the disease. The authors state that the eVOC anatomical terminology has the advantage of being simple and purely descriptive, without the interpretational bias that may be associated with functional annotation systems such as GO. Table 2 identifies the key objects and links in their approach.

The authors first identify the top ranked eVOC terms for a given disease (through L3). This is done by calculating a score that depends upon how frequently a term is associated with the disease in MEDLINE (L2), as well as upon how often the term is used to annotate RefSeq genes (L1). The former is an asymmetric weight calculated as the number of abstracts containing both the disease name and the eVOC term divided by the number of abstracts with the disease name. The later weight is also asymmetric and is calculated as the number of RefSeq genes annotated by the term divided by the number of annotated genes. Here annotation counts for an eVOC term include counts for descendent terms in the eVOC hierarchy. Finally the L3 weight between each eVOC anatomy term and disease name is calculated as $[2 * association\ weight + annotation\ weight]/2$. They then select the top scoring n eVOC terms as characterizing the disease. L5 which is the inferred link to new genes considers expression based annotation obtained from the Ensembl genomic database[9]. The final candidate gene list contains those annotated with at

Table 2 eVoC – Tiffin et al. (2005)

Type ID	Object Type	Object representation/link derivation	Notes
O1	Disease	Disease term (and retrieved MEDLINE set)	–
O2	eVoc term	eVoc term (and retrieved MEDLINE set)	–
O3	Gene	RefSeq entries	–
O4	Gene	Gene entries in Ensembl	–
L1	O2, O3	Frequency based annotation weight	A
L2	O1, O2	Frequency based association weight	A
L3	O1, O3	Score from L1 and L2	
L4	O1, O4	Expression of O3 in O2 tissues from Ensemble	L
L5	O1, O4	Inferred from L3 and L4	
IM	*Genes annotated with at least n – m of the n top ranked eVOC terms characterizing the disease*		

[8] http://www.sanbi.ac.za/evoc

[9] http://www.ensembl.org/Homo_sapiens/martview

least $n - m$ of the eVOC terms characterizing the disease. m functions as a slack parameter on the degree of matching. Optimal values for n and m were determined using training data and then evaluated on an independent test dataset.

3.3 BITOLA (Hristovski et al. 2001, 2003)

BIOTLA [13, 14] is a text mining system designed for the biomedical domain in general. We include it here as it has also been extended to identify novel genes for diseases. Using association rules with confidence and support scores, BITOLA builds on Swanson and Smalheiser's Open and Closed discovery strategies. In their earlier work [14], objects of interest are represented only by MeSH concepts. In later work [13], tailored to the gene–disease problem, they also consider gene names (that are not necessarily MeSH terms). Links are derived from co-occurrence data and are weighted by *confidence* and *support* scores. The *confidence* in a link between two MeSH concepts X and Y is an asymmetric weight which is the number of records having both X and Y divided by the number of records with X alone (or Y alone depending upon the perspective). Given a starting concept, X, associated Y concepts are found. Z concepts that are in turn associated with Y are identified. Each X–Z combination defines a novel relationship if they are not already directly associated. Filters may be applied to constrain the nature of the bridging Y concepts to those belonging to specific UMLS[10] semantic types of interest. Similarly links may be filtered based on threshold values for confidence or support. In the later version, tailored to the gene–disease application, they also provide filters to constrain the gene and disease to the same chromosomal region. They were able to postulate FLNA as a candidate gene for Bilateral perisylvian PMG, a malformation of cortical development in the brain [13]. More details about BITOLA are available in the chapter titled 'Literature-Based Knowledge Discovery using Natural Language Processing'. In Table 3 we show the key objects and links in their approach when

Table 3 BITOLA – Hristovski et al. (2003)

Type ID	Object type	Object representation/link derivation	Notes
O1	Disease	Disease MeSH term (and corresponding MEDLINE records with this term)	–
O2	Cell function	MeSH term of UMLS semantic type "cell function" (and corresponding MEDLINE records with this MeSH term)	–
O3	Gene	Gene names or aliases (and corresponding MEDLINE set with any of these terms)	–
L1	O1, O2	Confidence in O2 given O1	A
L2	O2, O3	Confidence in O3 given O2	A
L2	O1, O3	Inferred from L1 and L2 with added chromosomal constraint	A
IM	*Association rules exploiting transitivity with calculated confidence and support scores*		

[10] http://www.nlm.nih.gov/research/umls/

terms with the UMLS semantic type *cell function* are chosen as bridges between diseases and genes.

One observation to make at this point is that final scores on suggestions do not incorporate weights calculated for the intermediate links through the bridging Y pathways. Thus although a suggested Z concept may have a high confidence connection to its Y concept, this Y in turn may have a low confidence connection to the starting X concept (relative to other Y terms on the list).

3.4 Adamic et al. (2002)

In this work [2] the authors identify gene symbol occurrences (official and alias symbols) in MEDLINE records retrieved for a given disease. They then calculate the strength of the relationship between the gene and the disease by comparing the observed number of documents and the expected number of documents in which the gene is mentioned. Expected frequencies are determined assuming a random distribution of the gene term. They state that their work could be used to maintain a list of known genes for a disease. They do not explicitly explore "new" connections. However we include their research here as this approach of identifying genes that exhibit a statistically significant occurrence pattern in the disease literature is at the foundation of several papers and systems (Table 4).

One example is the newly formed Autoimmune Disease Database [16]. In it occurrences of gene names in document sets retrieved for diseases or disease names in document sets retrieved for genes are assessed for significance. LitMiner [18] also uses co-occurrence as the basis for relating two entities. Several types of entities are considered including genes and diseases. Unlike the Adamic et al. effort their link weight is symmetric and is calculated as the observed co-occurrence frequency divided by that expected by chance alone. MedGene[11] [15] is our last example of a system that relies on co-occurrence data. After comparing several statistical methods such as chi-square and Fishers exact probabilities, the authors select a symmetric measure called the natural log of the product of frequency. This is the product of two ratios. One is the disease–gene co-occurrence frequency divided by the disease frequency. The other is the disease–gene co-occurrence frequency divided by the gene frequency.

Table 4 Adamic et al. (2002)

Type ID	Object type	Object representation/link derivation	Notes
O1	Disease	Disease search terms	–
O2	Gene	Names and aliases	–
L1	O1, O2	Occurrence of O2 in MEDLINE records retrieved by O1	A
IM	*Comparison of observed and expected occurrences of O2 in O1*		

[11] http://hipseq.med.harvard.edu/MEDGENE/login.jsp

3.5 *Wilkinson and Huberman (2004)*

Wilkinson and Huberman [28] take the notion of significant co-occurrences and expand it to get additional capabilities. Specifically they explore methods for finding communities of genes that are likely to be functionally related in a given context such as that defined by a particular disease. Given a set of documents for a disease, they first identify occurrences of gene and protein names in each article. They then limit the identified gene set to those that are statistically relevant to the topic (disease). This is again done by comparing the observed number of gene occurrences with the expected number estimated, assuming no correlation between the gene and the disease. Specifically, given that two uncorrelated terms co-occur according to a binomial distribution, they consider observed gene-disease co-occurrences of at least one standard deviation greater than the binomial expected value as statistically relevant (Table 5).

They then create a co-occurrence network of the relevant genes and apply a graph partitioning algorithm to identify communities. Links in the network are not weighted and simply indicate that the genes co-occur. Their graph partitioning algorithm is based on the concept of the "betweenness centrality" of an edge. The betweenness of an edge AB is defined as the number of shortest paths between pairs of other vertices that contain AB. The edge with the highest betweenness is likely an intercommunity edge and is removed, thus breaking up the network into two or more connected components. This process iterates till certain stopping conditions are met. At each iteration the betweenness scores are recomputed. At the end, connected gene sets are declared to form a "community".

When applied to the disease topic 'colon cancer', the authors show that functionally unrelated genes tend to be placed in separate communities even if they exhibit some co-occurrence. Their method is offered as an approach for summarizing available information. The communities also indicate new directions for research based on connections among genes that may otherwise be overlooked or that would require much time and effort to be found manually. Their paper presents an analysis of select gene communities found for colon cancer. For example, they show that COX-1 and COX-2, isoforms of cyclooxygenases, are correctly placed in different communities as they are involved in different mechanisms. They also suggest possible connections between a set of phospholipase A2 genes and the gene FACL4 in the context of this disease.

Table 5 Wilkinson and Huberman (2004)

Type ID	Object type	Object representation/link derivation	Notes
O1	Disease	Disease terms (and corresponding MEDLINE records with these terms)	–
O2	Gene	Occurrences of names and aliases in O1 documents	–
L1	Set of O2	Membership in a common community	
IM	*Identify connected components by splitting the network using criteria based on "betweenness"*		

3.6 Analysis of Gene-Disease LBD Approaches

Objects: Given our selection of papers, the key objects are in all cases genes and diseases. However, we observe differences in representation across the studies. Representing a disease by its MeSH term or by its free-text terms can make a significant difference. For example, *alopecia areata* retrieves 1,768 records when limited to MeSH whereas it retrieves 17% more records (2,069) when searched without any limits. Also diseases are sometimes represented by their document set and sometimes by the documents sets corresponding to their pathological conditions. With genes we see an even greater variety in representation – from the occurrences of gene names and aliases in MEDLINE to gene sequence and gene expression data. Note also that the set of genes considered itself may vary. For example, G2D considers only those sequences that map to the disease chromosomal region while BITOLA allows this as an option.

Links: Even greater variability is seen in the types of links utilized. First, it is not surprising to see co-occurrence used for predicting disease–gene connections (Perez-Iratxeta et al., Adamic et al.) as co-occurrence is widely used in information extraction and text mining research. Examples include the efforts on predicting gene–gene relationships [18], transcription factor associations [19], and protein–protein interactions [7]. In each of the studies reviewed here some statistical assessment is undertaken to gauge significance of the proposed relationship. Typically this is some comparison of the observed co-occurrence frequency and the frequency expected assuming that the two objects are randomly paired.

Consider now the approach that takes advantage of intermediate conceptual bridges (links) such as through chemical terms (G2D), through eVOC (anatomical) terms and through terms representing cellular functions (BITOLA). In effect, these methods require particular varieties of semantics to tie the disease and the gene. An open question is how to determine the advantages gained by these semantic constraints when compared with co-occurrence based efforts. Likely false positives drop due to these requirements. However, is it the case that text mining, designed to moor on the fringe of the known, is better served by less constrained methods? Possibly this question can only be answered empirically.

Comparing the studies that use intermediate links also begs the question as to which type of connector is more effective. A point to note in this regard is that in G2D the disease and gene are at least four steps apart as its logic takes one from a disease to its pathological conditions to chemicals to GO annotations to RefSeq sequences to homologous sequences in a chromosomal region. Whereas in the eVOC approach only three steps are involved. Now is it the case that with every additional step there is an added risk of error? More fundamentally how do these different connectors, chemical and functional links as in G2D and anatomical links as in the eVOC system, compare? Could the GO cellular component vocabulary be used as effectively in G2D? Note that one could also use genes themselves as connectors between diseases and other genes. For example, in Chilibot [8] this strategy has been used to discover connections between phenomenon (such as long-term potentiation) and other genes. Genes as connectors are also implicit in the research by (Wilkinson and Huberman) on finding gene communities for diseases.

Clearly one direction of research is to explore more vocabularies as potential connectors. A related research direction is also to use vocabularies in combination which would require the study of evidence combination models. The choice of connector(s) could also be problem specific, i.e., depending on what is know about the particular disease. In that case a more general approach leaving the choice of conceptual bridge to the user as in BITOLA might be the best. This is another area where more research could be done. Moreover, as we accumulate experience with vocabularies for connectors, we might even begin to identify preferred characteristics (in addition to semantics). For instance, it may be that the more specific term subsets of a GO vocabulary are of greater value. Or, perhaps terms with high usage are more important. Observe that weights in the eVOC system are directly related to frequency of annotation with the term.

Inference methods: Differences in how measures exploit co-occurrence data are obvious but probably not as significant as other differences that may be observed. For example, inference methods that rely on a single path (BITOLA and G2D for links between GO terms and diseases) are categorically different from those that favour multiple paths. The eVOC system expects to find at least $m-n$ bridging anatomical concepts out of the n characterizing a disease. That is a tighter constraint. In a sense an extension of this notion is found in the research of Wilkinson and Huberman where the level of interest in a gene depends upon the 'community' to which it belongs.

A second major aspect to consider is one that is almost never addressed in text mining sytems. This aspect arises in the context of symmetric versus asymmetric methods. Given a system, can one expect to get consistent results whether we start from a disease seeking genes or we start from a gene seeking diseases? Take for example BITOLA that uses an asymmetric measure. Given the way in which confidence scores are computed, it is not clear if compatible results will be obtained. These scores are conditional probabilities that rely on the starting condition and the condition at each node of the path leading to the target. Thus directionality will matter. Of course, there is a natural perspective on a given problem, namely, the perspective of the user. However, it may be the case that for a given disease D, a gene G is identified as most interesting. Whereas from the perspective of the same gene, D may not be the most interesting disease. Perhaps one possible approach with asymmetric strategies would be to traverse both directions for a given problem and take the intersection of the top ranking suggestions.

Additional Knowledge Sources: While making explicit the additional sources used, our framework also suggests alternative designs. For example, the genes (O3 in Table 2) in the eVOC study could be represented by MEDLINE searches and L1 could be the co-occurrence in MEDLINE of eVOC terms and gene names. Assessing this strategy would at least tell us about the added value of using expression data from Ensemble. Alternatively, diseases could be represented by their description in

OMIM (and optionally also the linked documents) and the eVOC terms could be identified in these descriptions. As a final example, one could perhaps extend both the G2D system and the eVOC study to consider communities of genes where the edges between genes are drawn as a function of their common tissue expression patterns or their sequence similarities.

4 General Purpose Biomedical LBD Systems

We now consider LBD systems designed as general purpose biomedical text mining systems. Such systems are not tied to specific applications such as the discovery of gene–disease associations or protein–protein interactions. ARROWSMITH, BITOLA, IRIDESCENT, LitLinker and Manjal are some of the key domain independent systems available for use on the Web. In addition there are the research efforts of Weeber et al. [27] and Gordon and Lindsay [10], among others, which have extended and explored LBD strategies. Except for Manjal, the LBD systems listed above are (likely to be) described in detailed in other chapters of this book. Hence they will only be briefly reviewed here. In addition we present Manjal, a general purpose LBD system that we have developed at the University of Iowa.

BITOLA [13, 14] has also been described in the context of gene–disease links. As shown in Table 6 the objects of interest in BITOLA are topics represented by MeSH terms. The later version (2003) expands this to include genes as represented by their names and aliases. Open and Closed discovery are offered but the greater emphasis appears to be on using Open discovery to identify indirect relationships. BITOLA computes support and confidence from the association rules formalism to gauge the association strength between concept pairs. ARROWSMITH is likely the oldest general purpose LBD system implemented. There are presently two versions of ARROWSMITH, viz., University of Chicago version and University of Illinois – Chicago version. Both implement Closed discovery. The University of Chicago version allows one to upload two sets of retrieved MEDLINE records corresponding to two topics. These sets are then compared to find the list of intersecting title words, phrases and MeSH terms. These intersections are presented to the user as a ranked list where the ranking strategy also considers the common MeSH terms between the two starting query topics. The key difference in the University of Illinois – Chicago version is that the literature search step is integrated into the discovery system.

As described on their web site, LitLinker[12], considers objects represented by MeSH terms and implements Open discovery. According to the description in [21], they exploit correlations between terms calculated using the Apriori algorithm [3]. Finally there is the commercial system IRIDESCENT [29, 30]. It includes genes; diseases, disorders, syndromes or phenotypes; chemical compounds and small molecules; and drug names as objects. Although not truly a general purpose system, we include it here given its variety of objects and, we believe, its extensibility. Across a few papers they experiment with different probabilistic measures

[12] http://litlinker.ischool.washington.edu/

Table 6 General purpose biomedical LBD systems

Type ID	Object type	Object representation/link derivation	Notes
		BITOLA	
O1	Topic	MeSH concept and retrieved MEDLINE records	–
L1	Pairs of topics	Co-occurrence	A
L2	Pairs of topics	Implicit: connections through intermediate MeSH terms of specified semantic type	A
IM	*Open discovery with confidence and support weights*		
		ARROWSMITH (U. Chicago and U. Illinois – Chicago)	
O1	Topic	PubMed search and retrieved MEDLINE records	–
L1	Pairs of topics	Co-occurrence	
L2	Pairs of topics	Implicit: connections through shared title words, phrases, MeSH terms	
IM	*Closed discovery with frequency based weights*		
		LitLinker	
O1	Topic	MeSH term and retrieved MEDLINE records	–
L1	Pairs of topics	Co-occurrence	A
L2	Pairs of topics	Implicit: connections through intermediate MeSH terms of specified semantic type	A
IM	*Open discovery with weights calculated using support*		
		IRIDESCENT	
O1	Disease	OMIM entries and retrieved MEDLINE records	–
O2	Genes	Entrez Gene entries and retrieved MEDLINE records	–
O3	Chemicals	MeSH concepts and retrieved MEDLINE records	–
O4	Drugs	(from FDA) and retrieved MEDLINE records	–
O5	GO terms	(from GO) and retrieved MEDLINE records	–
L1	Pairs of objects	Co-occurrence	
L2	Pairs of objects	Implicit: connections through other objects	
IM	*Open discovery with probabilistic weights*		
		Manjal	
O1	Topic	PubMed search and topic profile from retrieved MEDLINE records	–
O2	Topic	MeSH concept and topic profile from retrieved MEDLINE records	–
L1	Pairs of topics	Co-occurrence	
L1	Pairs of topics	Profile similarity	
L3	Pairs of topics	Implicit: related through other topics	A
IM	*Open discovery, closed discovery, multi-topic analysis, bipartite topic analysis TFIDF weights*		

of association that may be used to gauge the relatedness between a pair. In [30], for example, they study mutual information and extend these in two ways to assess the value of implicit relationships identified using an Open discovery model. The extensions consider the different pathways connecting the two objects. Using IRIDESCENT they found, for example, that the drug chlorpromazine, which is normally used to treat problems such as psychotic disorders and also severe hiccups, would also reduce the progression of cardiac hypertrophy.

4.1 Manjal

Manjal, available on the Web[13] offers a variety of LBD options for mining MEDLINE. In each option a user specifies one or more topics, where a topic is any valid PubMed search. For each input topic provided, Manjal first retrieves records from MEDLINE after which it builds its profile, which is simply a weighted vector of terms. Terms are from the MeSH or/and RN fields of the records. Profile terms are assigned TF*IDF weights with cosine normalization. All the text mining functions offered in Manjal operate on top of these profiles. In essence, these functions can make use of similarities calculated between topic profiles, employ MeSH terms and their profiles as bridges between topics and relate topics based on shared MeSH terms. In addition Manjal also offers co-occurrence based analysis.

Manjal users may conduct Open discovery runs starting with a single topic. The end result is a ranked list of MeSH terms, organized by semantic type. Each ranked MeSH term represents a topic that might have an interesting (and implicit) connection to the starting topic. Bridging topics are also presented. In the Closed discovery option two topics are provided as input and their MeSH profiles are created. The MeSH terms they share and their combined weights provide the tentative bridges between the two starting topics. A third function extends the notion of the two-topic Closed discovery function to work with larger sets of topics. Profiles are built for each topic and the neighborhood of any given topic may be explored. Neighborhoods may be selected on the basis of profile similarity or on the basis of co-occurrence frequency. Manjal's user interface is graphical and interactive. Both nodes and links may be clicked to obtain further details including for example, the corresponding set of PubMed documents.

An upgraded beta version of Manjal (not public, access available by request) offers additional functions. For example, it allows analysis of bipartite topic sets. This is appropriate when the problem naturally breaks down into two groups of topics. For example, the two sets of topics could be a set of diseases and a set of genes, or a set of environmental toxins and a set of diseases etc. Using this function one may for example, rank members of one set in terms of their association with members of the other set.

In all of the above functions the user may constrain the text mining process by specifying the types of connections desired. Indeed it is desirable to do so as otherwise the process could generate an overwhelming amount of information. This is done by allowing only terms from certain UMLS semantic types to participate in the process. For example, in Open discovery the intermediate terms may be restricted to those of type *Cell Function* or *Gene or Genome.*

Manjal has tested successfully on a set of "benchmark" LBD problems that derive from the research of Swanson and Smalheiser [23]. This replication study is the most extensive performed thus far. Manjal has also been used to propose a beneficial relationship between the dietary substance *Curcumin Longa* also known as turmeric and disorders such as retinal diseases, Crohn's disease and problems of the spinal

[13] Manjal: http://sulu.info-science.uiowa.edu/Manjal.html

cord [24]. The postulated connections were through biochemical pathways involving several genes such as inflammatory genes. Interestingly, a recent pilot study [12] has been published where a pure curcumin preparation was administered to patients with ulcerative proctitis and patients with Crohn's disease. The authors conclude that the results encourage follow-up double-blind placebo-controlled studies.

4.2 Analysis of the General Purpose LBD Systems

Objects: We observe significant differences in how topics are conceptualized in these systems. With ARROWSMITH and Manjal users may start with any search that is legitimate in the PubMed system. However, the difference between these two is that in Manjal intermediate topics are defined by MeSH terms whereas with ARROWSMITH these are terms from the free-text fields. The remaining systems either limit themselves to MeSH for topic specification or to predefined objects whose names (or aliases) appear somewhere in the MEDLINE records. These differences are fundamental. For example, when the user is constrained to MeSH as input, complex queries such as *erythromycin AND antihistamines AND hypertension* cannot be considered. More generally, the space of input topics is unbounded with ARROWSMITH and Manjal. Whereas, with BITOLA and LitLinker, these are bounded by the MeSH vocabulary. A possible extension that remains consistent with the parameters of these systems is to allow for combinations of MeSH concepts as input topics. This would certainly remove some of the constraints, albeit at the cost of having to calculate various frequency based statistics at run time.

Links: As seen in LBD systems exploring gene–disease connections, there are differences in the way co-occurrence is used (or not) to define links. However, what is more interesting is the remarkable absence of "semantic" links. For example, although IRIDESCENT identifies disease sets, gene sets etc. from curated resources, it appears to ignore the links between the two object types available from say OMIM. The larger question to address concerns the extent to which these LBD systems may benefit from the inclusion of expert acknowledged relationships as available in curated databases. One option may be to utilize known relationships harvested from sources such as Entrez Gene and OMIM to build a network of associated objects. This could then be the basis of Open and Closed discovery algorithms. A second option could be to use a hybrid approach that allows one to smoothly incorporate both semantic relationships along with co-occurrence based information. Exploring evidence combination models is then an important research direction.

Points raised earlier about identifying implicit relationships from single paths versus multiple paths also apply here. IRIDESCENT, for example, probabilistically assesses the strength of the association between the target topic (in Open discovery) and the collection of intermediate topics connecting to the starting topic. Manjal also calculates a weight that is a function of the number and importance rating of intermediate paths. In contrast BITOLA, for example, offer single link paths. Their relative merits remain unknown though at the intuitive level strategies favouring multiple paths may be more reliable.

Inference Methods: The final aspect to (re-)consider is that of symmetric versus asymmetric inferencing strategies. Again, it is not clear as to the role of direction in all of these systems. Whether one starts with a gene topic looking for novel diseases or vice versa, it is unclear if compatible results are obtained. This too remains an open research area in the context of these general purpose LBD systems.

5 Predicting Relationships from the Web

We now turn our attention to efforts on discovering novel links from the Web. The main emphasis has been on discovering connections between people. However, some effort has been devoted to finding connections involving companies and industries as also between web pages.

5.1 Adamic and Adar (2003)

Adamic and Adar [1] aim to predict relationships between students at MIT and Stanford based on the similarity in characteristics extracted from their home pages. Specifically, the authors use the text, hyperlinks (inlinks and outlinks), and mailing list subscriptions on the web pages to "profile" students. Each individual feature is weighted by the inverse log of its frequency. Profile similarity is computed as the sum of the weights of the features in common. The authors also analyze predictions based on the individual feature types and find that the text of the home pages acts as the best predictor of a relationship. Table 7 represents the key objects and links.

Table 7 Adamic and Adar (2003)

Type ID	Object type	Object representation/link derivation	Notes
O1	Students at Stanford	(i) Text words in Home Page (ii) Inlinks in Home Page (iii) Outlinks in Home Page (iv) Mailing lists in Home Page (v) Composite of above	–
O2	Students at MIT	(i) Text words in Home Page (ii) Inlinks in Home Page (iii) Outlinks in Home Page (iv) Mailing lists in Home Page (v) Composite of above	–
L1	O1, O2	Sum of weights of features in common	
IM	*Similarity in profiles*		

Table 8 BenDov et al. (2004)

Type ID	Object type	Object representation/link derivation	Notes
O1	Person	Name and retrieved news items as identified by ClearForest software	–
L1	Pairs of people	Co-occurrence in sentence	U
L2	Pairs of people	(from L1) Implicitly via other people	U
IM	*Transitive relationships*		

Further analysis of the predictions made for the Stanford students can be found online[14]. Note that this research does not rely on co-occurrence as the home pages of two individuals are unlikely to overlap. We regard this research as LBD albeit working off non-traditional "documents".

5.2 Ben-Dov et al. (2004)

Working off approximately 9,100 documents from four news sites: CNN, BBC, CBS and Yahoo, the authors of this paper [5] identify novel relationships between person entities. Two entities are explicitly connected if they co-occur in a sentence. Two entities are implicitly connected if they form part of a transitive structure with an intermediate entity. For example they connect Osama Bin Laden and Pope John Paul via Ramzi Yousef. Bin Laden is connected to Yousef in several ways. For example one document mentions that Yousef stayed in Bin Laden's house. Two documents mention a book by Simon Reeve called "The New Jackals: Ramzi Yousef, Osama bin Laden and Future of Terrorism". Yousef is connected to the Pope by reports on an attempted assassination. They identify entities in the news articles using an information extraction tool (ClearForest[15]). They also extract semantic links between entities using NLP-based methods such as by identifying patterns involving noun phrases, verbs, etc. However, they do not use semantic links for knowledge discovery. Table 8 shows the details.

5.3 Cory (1997)

Although the research described in this paper [9] addresses the humanities domain in general, the author also focuses on relationships between people. This work is a direct application of Swanson's Open discovery approach to humanities data obtained

[14] http://www.hpl.hp.com/research/idl/papers/web10/frequency.html

[15] http://www.clearforest.com/

Table 9 Cory (1997)

Type ID	Object type	Object representation/link derivation	Notes
O1	Writer	Name and retrieved records from humanities index	–
L1	Pairs of writers	Co-occurrence	U
L2	Pairs of writers	(from L1) Implicitly linked via other writers	U
IM	*Transitive relationships*		

from the humanities index (under the WILS database)[16]. The aim here is to identify new analogies. The author says that this application domain is difficult because the language of humanities articles is not as structured and formal as that of medical articles. The author notes that person names fits the bill and uses those. Starting from a person (A) Cory retrieves documents from the humanities index and then identifies people names (Bs) in the titles other than A itself. Semantics for the relations between the Bs and A are manually established from the documents and interesting Bs identified. For each of these Bs, Cory conducts fresh searches and identifies names (Cs) in the titles of the new articles retrieved and again relation semantics are manually established. Then via a transitive analysis the author connects A with interesting Cs. A relationship is novel if a query containing both A and C does not retrieve any documents (Table 9). Cory finds a novel analogy for the twentieth century writer Robert Frost in the form of a classical second century BCE Greek writer, Carneades, via a nineteenth century writer, William James (1842–1910).

5.4 Kumar et al. (1999)

In this paper [17] the authors describe an approach to identify online communities. They concentrate on "new" communities that are not yet established or are implicitly defined. By this they mean communities at a finer level of detail such as the community of turkish student organizations in the US. These communities are typically not yet listed on any web portal. Their operational definition of a community is a densely connected bipartite subgraph known as a 'core'. A core consists of fans that are pages with outlinks and centers that are pages with inlinks. Fans can be thought of as specialized hubs and centers are the pages with the required information. The authors define an iterative procedure that consists of many pruning steps. Starting from a large set of nodes they keep pruning until they identify a community (or core) in which both the fans and centers have a minimum number of outlinks and inlinks respectively. Using data crawled by Alexa, consisting of over 200 million web pages, they identify communities such as the Australian Fire Brigade Services. They also explore temporal analysis to verify how many communities,

[16] We acknowledge that we have stretched our definition of Web based LBD works to include this WILS database research. We do this given the innovativeness of the work and its direct use of LBD.

out of a random sample of 400 communities identified, survive for more than 18 months. Interestingly they found that approximately 70% of the communities were still alive (Table 10).

5.5 Tan and Kumar (2001)

Tan and Kumar [25] propose an approach to identify novel links between web pages from sequences in user session data. The aim is to help in restructuring web sites so as to better conform to the navigational behavior of users. They do this by identifying sequential and non-sequential indirect associations from user session data. They first identify all the frequent itemsets from the data (using algorithms such as the Apriori algorithm [3]). In the non-sequential case they postulate indirect associations between pairs of unrelated pages if they share intermediate sets of pages, called Mediator sets, that are frequently associated with them. In the sequential case they identify intermediate sequences, called mediator sequences, and infer indirect connections between pairs of unrelated pages that share mediator sequences via three kinds of inference mechanisms, viz., convergence, divergence and transitivity. These indirect connections suggest more optimal ways to structure web sites (Table 11).

5.6 Bernstein et al. (2002)

Bernstein et al. [6] address the goal of identifying relationships between companies, more specifically similarities, using a large corpus of business news. They combine information extraction techniques with network analysis and statistical approaches

Table 10 Kumar et al. (1999)

Type ID	Object type	Object representation/link derivation	Notes
O1	Web pages	URL address	–
L1	Pair of O1 objects	URL based connections	U
L2	Community of O1 objects	(from L1) ‘Cores’ with particular features	U
IM	*Presence in ‘core’ after pruning*		

Table 11 Tan and Kumar (2001)

Type ID	Object type	Object representation/link derivation	Notes
O1	Web page	URL	–
L1	Pair of O1 objects	Support based on co-occurrence frequency	
L2	Pair of O1 objects	Implicit: through intermediate sets	
L3	Pair of O1 objects	Implicit: through intermediate sequences	A
IM	*Convergence, divergence, and transitivity*		

Table 12 Bernstein et al. (2002)

Type ID	Object type	Object representation/link derivation	Notes
O1	Company	Company name and vector of co-occurring companies	–
O2	Industry	Average of company vectors	–
L1	Pairs of O1 objects	Similarity of vectors	
L2	O1, O2	Similarity of vectors	
L3	Pairs of O2 objects	Similarity of vectors	
IM	*Cosine similarity*		

to extract knowledge of company interrelationships. Distinct company names are identified in a news collection from a 4-month period. Companies, represented as nodes, are displayed in a co-occurrence network to provide a visual overview of their distribution and connectivity. Going further, they represent each company by a vector of its co-occurring companies and calculate cosine similarities between vectors. Note that in this approach two companies may not co-occur and yet show high similarity. They also use the same principles to explore the relationship between individual companies and different industries as well as between industries. An industry is regarded as a cluster of companies. An industry vector is defined as the average of the vectors of the companies that belong to it. Table 12 abstracts from their work the key features of their methods.

5.7 Analysis of Web Based LBD

The key objects considered are students, web pages, companies and industries. Students for example, were represented by their home pages optionally augmented with their inlinks and/or their outlinks. Several alternative representations may be considered. One could use their entries in blogs, the set of papers presented as seen in conference web sites or the web pages of related individuals such as professors and co-authors. Each variety of representation provides a different perspective on the individual student that may be useful in determining novel relationships.

With links we see an interesting variety, indicative of the broad potential with the Web. For example, in addition to exploiting URL-based links, we see relationships inferred from user-access data for Web sites. Similarly, one can imagine search logs being a good source of implicit relations between pages, web sites, products, organizations and possibly also between web users. However, the anonymous nature of search logs makes the detection of user connections difficult. With companies we see that in addition to generating a co-occurrence based network Bernstein et al. uses a second-order strategy that groups companies by calculating similarities based on company co-occurrence feature vectors. This allows for two companies to be very similar without co-occurring.

Beyond these object and link specific points, several general observations may be made. The first is that there is little research on the web involving methods that are the same or analogous to LBD. This is particularly striking when compared to the level of activity in the biomedical domain. There are several potential explanations for this discrepancy. First, the kinds of goals that may be targeted are not as easily specifiable on the web as in biomedicine. In biomedicine, especially in bioinformatics, the key entities are widely understood to include genes, diseases, proteins, chemicals etc. This understanding is reflected in the kinds of curated resources that have been created. Thus to look for novel associations, such as those between diseases and genes, or drugs and genes follows naturally from the set of key entities. Despite the heterogeneity of the Web, very few key entity types have been explored for LBD.

A second possible reason for the paucity of LBD research in the web domain is the ambiguity challenge. For example, a straightforward application of Open discovery would involve starting with a search on an input topic. Even when focussed on a person as the input topic, we will need to filter the retrieved set in order to disambiguate between the multiple individuals likely to share the same name. We note that this sub-problem is itself being directly addressed (e.g., [11]). Although ambiguity resolution is also required in biomedical LBD, the problem is far more pronounced on the Web, given its heterogeneity and especially given its much faster pace of growth.

A third possible explanation for the lack of LBD research on the Web is a very practical one, which is the non-availability of appropriate datasets. Researchers in the biomedical domain may easily avail of the MEDLINE database, PubMed APIs and the UMLS vocabularies. This has created an incredibly hospitable environment for LBD research. Added to this are the many curated resources such as Entrez Gene and OMIM, typically with an option for data downloads. In contrast, the Web is close to being inhospitable to LBD research. For example, API's to search systems such as Google or Yahoo! limit users to only 1,000 and 5,000 daily searches, respectively. Moreover, each search is limited to the top few results. Also these APIs do not provide all the search options that are implemented on the respective web sites. For example, the Google API does not allow blog search. Avoiding these search systems implies the need to crawl the web and develop ones own Web datasets. Collections such as Alexa crawls[17], available at a fairly low cost, are certainly in the right direction. But the real power of LBD is in identifying *novel* hypotheses which implies working with information that is *current*. Thus although working off predefined collections may help in refining methods, it is unlikely to be of real value to end users.

To counter these challenges, the web, with information on almost every type of entity, offers excellent opportunities for existing LBD methods. Consider the kinds of problems addressed in the papers reviewed. A key emphasis is on finding implicit relations between people: students from two universities (Adamic and Adar), authors across time (Cory), and individuals mentioned in news articles

[17] http://www.amazon.com/b/ref=sc_fe_c_0_239513011_1/103-4334540-2295806ie=UTF8& node=12782661& no=239513011&me=A36L942TSJ2AJA

(Ben-Dov et al.). There is also some emphasis on individual pages as key entities in terms of defining emerging communities of web pages and toward reshaping websites. There is surprisingly little work on companies. However, these are just the beginnings in terms of LBD on the Web. For example, even within the space of individuals, we have an open research forum given the different classifications of professions, affiliations etc.

The Web also offers excellent opportunities for developing new LBD methods. In fact, the development of new methods is almost inevitable given that each document (web page) is readily characterized not only by its content but also by its inlinks and outlinks. Interestingly the methods proposed by Tan and Kumar and by Kumar et al. consider mainly the URLs and links. It may be the case, for example, that by considering page content based similarities as well, more cohesive cores are identified by the latter's method. This is also somewhat indicated by the Adamic and Adar research, where the text of the home page is determined to be the best predictor of a relationship. LBD methods exploiting URLs may also contribute to biomedical LBD given the availability of fast growing full-text collections such as PubMed Central. Thus citations to and from biomedical articles may eventually be exploited for LBD.

6 Conclusions

We presented an overview of literature-based discovery methods using a common framework for analysis. The framework focusses on the key objects, links, inference methods and knowledge sources used. The analysis was presented in three parts. The first part was constrained to a single theme of finding novel gene–disease connections. In the second part we analyzed general purpose LBD systems in biomedicine. The third part analyzed the few papers that use LBD or analogous methods on the Web.

The framework allowed us to perform a focussed comparison and analysis of LBD methods. In the process several open questions and directions for research were identified. For example, in the gene–disease context an important question is on the relative merits of single link discovery paths versus multilink paths. Another important angle for research is on the design of evidence-combination models that consider multiple intermediate vocabularies. With general-purpose biomedical LBD systems an example open research direction is on incorporating semantic links from curated databases into the process. Links of interest include not only those extracted from texts but also those readily available in curated resources. Despite the prevalence of LBD research in biomedicine we still do not know the relative merits of implicit connections that are co-occurrence based versus those derived from more semantic/conceptual links. Also needed is research studying the implications of symmetric versus asymmetric LBD strategies. We believe that this question has been given little or no attention in the literature. As a consequence, there is the risk of underrating or overrating a hypothesis given the chosen direction of the LBD analysis. This chapter also compares LBD on the web with LBD in biomedicine. It is clear that LBD on the web is at a very early stage. However, LBD opportunities are abundant, especially if we can cross a few of the key hurdles. Moreover, methods such as URL-based LBD strategies, developed on the Web have the potential to influence methods for biomedicine.

There are several limitations of the analysis presented in this chapter. As said initially this chapter is not a comprehensive review of LBD research. Thus for example, we ignored interesting problems such as identifying implicit drug–disease, protein–protein interactions. In the general-purpose LBD research, we reviewed only LBD systems as opposed to papers that presented strategies without having a freely accessible system. Also we did not focus on the types of experiments and the results obtained in each paper. Instead we considered primarily the key methodological details.

To conclude, our framework-based review provides a better understanding of the similarities and differences across LBD systems and methods. Through this endeavor, our own knowledge on the evolution of LBD research in different domains and some of the key hurdles has greatly improved. This chapter also raises several questions and identifies avenues for extending LBD research. Hopefully these will guide the efforts of the LBD research and development community.

Acknowledgements This material is partly based upon work supported by the National Science Foundation under Grant No. 0312356 award to Srinivasan. Any opinions, findings, and conclusions or recommendations expressed in this material are those of the author(s) and do not necessarily reflect the views of the National Science Foundation.

References

1. Lada A. Adamic and Eytan Adar. Friends and neighbors on the web. *Social Networks*, 25(3):211–230, 2003
2. Lada A. Adamic, Dennis Wilkinson, Bernardo A. Huberman, and Eytan Adar. A Literature Based Method for Identifying Gene-Disease Connections. In *Proceedings of the IEEE Computer Society Bioinformatics Conference (CSB 2002)*, pp. 109–117, 2002
3. Rakesh Agrawal and Ramakrishnan Srikant. Mining Sequential Patterns. In *Proceedings of the Eleventh International Conference on Data Engineering*, pp. 3–14, 1995
4. Michael Ashburner, Catherine A. Ball, Judith A. Blake, David Botstein, Heather Butler, J Micheal Cherry, Allan P. Davis, Kara Dolinski, Selina S. Dwight, Janan T. Eppig, Midori A. Harris, David P. Hill, Laurie Issel-Tarver, Andrew Kasarskis, Suzanna Lewis, John C. Matese, Joel E. Richardson, Martin Ringwald, Gerald M. Rubin, and Gavin Sherlock. Gene ontology: tool for the unification of biology. The Gene Ontology Consortium. *Nature Genetics*, 25:2529, 2000
5. Moty Ben-Dov, Wendy Wu, Ronan Feldman, and Paul A. Cairns. Improving Knowledge Discovery by Combining Text-Mining and Link-Analysis Techniques. In *Proceedings of the SIAM International Conference on Data Mining*, 2004
6. Abraham Bernstein, Scott Clearwater, Shawndra Hill, Claudia Perlich, and Foster Provost. Discovering Knowledge from Relational Data Extracted from Business News. In *Proceedings of Workshop on Multi-Relational Data Mining (MRDM 2002)*, 2002

7. Peter M. Bowers, Matteo Pellegrini, Mike J. Thompson, Joe Fierro, Todd O. Yeates, and David Eisenberg. Prolinks: a database of protein functional linkages derived from coevolution. *Genome Biology*, 5(R35), 2004
8. Hao Chen and Burt M. Sharp. Content-rich biological network constructed by mining PubMed abstracts. *BMC Bioinformatics*, 5(147), 2004
9. Kenneth A. Cory. Discovering hidden analogies in an online humanities database. *Computers and the Humanities*, 31:1–12, 1997
10. Michael D. Gordon, Robert K. Lindsay, and Weiguo Fan. Literature-based discovery on the World Wide Web. *ACM Transactions on Internet Technologies (TOIT)*, 2(4):261–275, 2002
11. Ramanathan V. Guha and A. Garg. Disambiguating People in Search. Technical Report, Stanford University, 2004
12. Peter R. Holt, Seymour Katz, and Robert Kirshoff. Curcumin therapy in inflammatory bowel disease: a pilot study. *Digestive Diseases and Sciences*, 50(11):2191–2193, 2005
13. Dimitar Hristovski, Borut Peterlin, Joyce A. Mitchell, and Susanne M. Humphrey. Improving literature based discovery support by genetic knowledge integration. *Studies in Health Technology and Informatics*, 95:68–73, 2003
14. Dimitar Hristovski, Janez Stare, Borut Peterlin, and Saso Dzeroski. Supporting discovery in medicine by association rule mining in medline and UMLS. *Medinfo*, 10(Pt 2):1344–1348, 2001
15. Yanhui Hu, Lisa M. Hines, Haifeng Weng, Dongmei Zuo, Miguel Rivera, Andrea Richardson, and Joshua LaBaer. Analysis of genomic and proteomic data using advanced literature mining. *Journal of Proteome Research*, 2:405–12, 2003
16. Thomas Karopka, Juliane Fluck, Heinz-Theodor Mevissen, and Änne Glass. The Autoimmune Disease Database: a dynamically compiled literature-derived database. *BMC Bioinformatics*, 7(325), 2006
17. Ravi Kumar, Prabhakar Raghavan, Sridhar Rajagopalan, and Andrew Tomkins. Trawling the Web for Emerging Cyber-Communities. In *Proceedings of the Eighth International Conference on World Wide Web (WWW-8)*, pp. 1481–1493, 1999
18. Holger Maier, Stefanie Döhr, Korbinian Grote, Sean O'Keeffe, Thomas, Werner, Martin Hrabé de Angelis, and Ralf Schneider. LitMiner and WikiGene: identifying problem-related key players of gene regulation using publication abstracts. *Nucleic Acids Research*, 33:W779–W782, 2005
19. Hong Pan, Li Zuo, Vidhu Choudhary, Zhuo Zhang, Shoi H. Leow, Fui T. Chong, Yingliang Huang, Victor W.S. Ong, Bijayalaxmi Mohanty, Sin L. Tan, S.P.T. Krishnan, and Vladimir B. Bajic. Dragon TF Association Miner: a system for exploring transcription factor associations through text-mining. *Nucleic Acids Research*, 32:W230–W234, 2004
20. Carolina Perez-Iratxeta, Peer Bork, and Miguel A. Andrade. Association of genes to genetically inherited diseases using data mining. *Nature Genetics*, 31(3):316–319, 2002
21. Wanda Pratt and Meliha Yetisgen-Yildiz. Litlinker: Capturing Connections Across the Biomedical Literature. In *Proceedings of the International Conference on Knowledge Capture (K-CAP 2003)*, pp. 105–112, 2003
22. Aditya K. Sehgal, Xing Y. Qiu, and Padmini Srinivasan. Mining MEDLINE Metadata to Explore Genes and their Connections. In *Proceedings of the 2003 SIGIR Workshop on Text Analysis and Search for Bioinformatics*, 2003
23. Padmini Srinivasan. Text mining: generating hypotheses from MEDLINE. *Journal of the American Society for Information Science and Technology (JASIST)*, 55(5):396–413, 2004
24. Padmini Srinivasan and Bisharah Libbus. Mining MEDLINE for implicit links between dietary substances and diseases. *Bioinformatics*, Suppl. 1:I290–I296, 2004
25. Pang-Ning Tan and Vipin Kumar. Mining Indirect Associations in Web Data. In *Proceedings of the Workshop on Mining Logdata Across All Customer Touchpoints (WEBKDD '01)*, pp. 145–166, 2001
26. Nicki Tiffin, Janet F. Kelso, Alan R. Powell, Hong Pan, Vladimir B. Bajic, and Winston A. Hide. Integration of text- and data-mining using ontologies successfully selects disease gene candidates. *Nucleic Acids Research*, 33(5):1544–1552, 2005

27. Marc Weeber, Jan A. Kors, and Barend Mons. Online tools to support literature-based discovery in the life sciences. *Briefings in Bioinformatics*, 6(3):277–286, 2005
28. Dennis M. Wilkinson and Bernardo A. Huberman. A Method for Finding Communities of Related Genes. In *Proceedings of the National Academy of Sciences of the United States of America*, 101:5241–5248, 2004
29. Jonathan D. Wren. *The IRIDESCENT System: An Automated Data-Mining Method to Identify, Evaluate, and Analyze Sets of Relationships Within Textual Databases*. PhD thesis, University of Texas Southwestern Medical Center, 2003
30. Jonathan D. Wren, Raffi Bekeredjian, Jelena A. Stewart, Ralph V. Shohet, and Harold R. Garner. Knowledge discovery by automated identification and ranking of implicit relationships. *Bioinformatics*, 20(3):389–398, 2004

Evaluation of Literature-Based Discovery Systems

M. Yetisgen-Yildiz and W. Pratt

Abstract Evaluating discovery systems is a fundamentally challenging task because if they are successful, by definition they are capturing new knowledge that has yet to be proven useful. To overcome this difficulty, many researchers in literature-based discovery (LBD) replicated Swanson's discoveries to evaluate the performance of their systems. They reported overall success if one of the discoveries generated by their system was the same as Swanson's discovery. This type of evaluation is powerful yet incomplete because it does not inform us about the quality of the rest of the discoveries identified by the system nor does it test the generalizability of the results. Recently, alternative evaluation methods have been designed to provide more information on the overall performance of the systems. The purpose of this chapter is to review and analyze the current evaluation methods for LBD systems and to discuss potential ways to use these evaluation methods for comparing performance of different systems, rather than reporting the performance of only one system. We will also summarize the current approaches used to evaluate the graphical user interfaces of LBD systems.

1 Introduction

Evaluation plays an important role in the development of new fields such as literature-based discovery (LBD). Evaluation encourages scientific progress by supporting a systematic comparison of different techniques applied to a common

M. Yetisgen-Yildiz
The Information School, University of Washington, Seattle, USA
melihay@u.washington.edu

W. Pratt
The Information School, University of Washington, Seattle, USA
and
Biomedical and Health Informatics, School of Medicine, University of Washington, Seattle, USA
wpratt@u.washington.edu

P. Bruza and M. Weeber (eds.), *Literature-based Discovery,*
Springer Series in Information Science and Knowledge Management 15.

problem and allowing researchers to learn from each other's successes and failures. In this chapter, we will give an overview of the current state of evaluation in literature-based discovery research and discuss potential ways for future evaluations.

2 Evaluation Metrics

When developing an LBD system, it is critical to know how reliable the results are likely to be. Measuring the reliability of a prediction system requires two main components: a gold standard and an evaluation metric to measure the system's performance with respect to the gold standard. For a given starting term, which Swanson called *C-Term*, a typical LBD system produces two sets of terms; *linking terms* and *target terms*. The linking terms, which Swanson called *B-Terms*, directly connect a given starting term to the target terms, which Swanson called *A-Terms*. The gold standards used to evaluate those two sets of terms are different from each other, and the gold standard creation methods depend on which of the evaluation methods listed in Sect. 3 is used. We will describe how the gold standards for linking/target terms are created for certain types of evaluation methods in Sect. 3. For now, we will define the gold standards for linking/target terms as the two sets of terms that are known to be directly/indirectly connected to a given starting term. In this section, we will summarize the metrics used to measure the performance of LBD systems.

2.1 Information Retrieval Metrics

The main purpose of evaluation in information retrieval research (IR) is to measure IR systems' performance in returning the relevant documents and in not returning the non-relevant documents to user queries. In IR evaluation, the gold standard is the set of relevant documents and two most popular IR metrics used to measure system performance are *precision* and *recall* [1]. For a given query and an IR system, precision can be defined as the proportion of relevant documents in the set of documents returned by the system and recall can be defined as the proportion of the relevant documents retrieved by the system from the gold standard.

In contrast to IR systems, LBD systems return terms instead of documents. Thus, precision and recall are mainly used to measure the effectiveness of an LBD system in returning linking and target terms for a given starting term, rather than the effectiveness of an IR system in returning documents for a given query. Precision and recall for the LBD system evaluation are calculated with the following formulas:

$$\text{Precision} : P_i = \frac{\|T_i \cap G_i\|}{\|T_i\|}. \tag{1}$$

$$\text{Recall}: R_i = \frac{\|T_i \cap G_i\|}{\|G_i\|}, \tag{2}$$

where T_i is the set of linking/target terms generated by the LBD system for the starting term i, and G_i is the set of terms in the linking/target term gold standard that the LBD system created for the starting term i.

As with IR system evaluation, one challenge in interpreting precision and recall is that there is a trade-off between the two metrics. Usually a system that aims to achieve high precision will result in low recall and vice versa. To solve this problem, some information retrieval researchers invented a new measure called *F-Measure* which is a combined version of precision and recall. F-Measure is calculated with the following formula:

$$\text{F-Measure}: F = \frac{(1+\beta^2) \times R \times P}{(\beta^2 \times P) + R}, \tag{3}$$

where R is the recall, P is the precision, and β is the relative value of the precision. The most commonly used case $\beta = 1$ assigns equal emphasis on precision and recall, whereas a lower value assigns a higher emphasis on precision and a higher value assigns a higher emphasis on recall.

Another common method to combine precision and recall is to draw a precision-recall curve. In this curve, the x-axis corresponds to recall and the y-axis corresponds to precision. Because of the trade-off between precision and recall, precision-recall graphs usually have a concave shape. Trying to increase recall typically introduces more false positives (target terms that are not in the gold standard), and thereby reduces precision. Trying to increase precision typically reduces recall by decreasing the number of true positives (target terms that are in the gold standard). An ideal goal of a prediction system is to increase both precision and recall by making improvements to the system. In other words, the entire curve must move up and out to the right so that both recall and precision are higher at every point along the curve. The most common use of precision-recall curves is for system comparisons.

2.2 Receiver Operating Characteristics (ROC) Curve

Receiver Operating Characteristics (ROC) curve provides a graphical representation of the relationship between the true positive and false positive rate of a prediction system [2]. These curves are used frequently in comparing the effectiveness of different medical diagnostic tests. The y-axis corresponds to the *sensitivity* of the system. Sensitivity measures the performance of the system in predicting the true positives. The x-axis corresponds to the *specificity* (expressed as *1-specificity* in the graph). Specificity represents the ability of the system in identifying true negatives. The sensitivity and the specificity of a LBD system can be calculated as:

$$\text{Sensitivity}: Y_i = \frac{TP_i}{TP_i + FN_i}. \tag{4}$$

$$\text{1 - Specificity}: X_i = 1 - \frac{TN_i}{TN_i + FP_i}, \tag{5}$$

where for the starting term i; TP_i is the number of true positives (the target terms that are in the gold standard), FN_i is the number of false negatives (the gold standard terms that are not identified as target terms), FP_i is the number of false positives (the target terms that are not in the gold standard), and TN_i is the number of true negatives (the terms that are both not selected as target terms and not in the gold standard).

The ROC curves show the performance as a trade off between specificity and sensitivity of the prediction system. The area under the ROC is a convenient way of comparing different prediction systems. A random system has an area of 0.5, while and ideal one has an area of 1.

2.3 Probabilistic Approaches

Because the purpose of LBD systems is to predict novel connections between medical terms, it is also important to compare their prediction performance with that of pure random prediction. One way to accomplish this objective is to calculate the probability of randomly achieving the performance of a given LBD system. This probability can be modeled with *hypergeometric distribution*. Suppose for a given starting term, an LBD system returns k target terms where i of the target terms that are in the gold standard, there are n terms in the gold standard and there are m terms in the search space of the system. The probability of having i gold standard terms in randomly selected k target terms is calculated with the following formula:

$$p(x = i) = \frac{\binom{n}{i}\binom{m-n}{k-i}}{\binom{m}{k}}. \tag{6}$$

If the value of p is close to zero, achieving the performance of the LBD system by randomly selecting the target terms is highly unlikely. If the value of p is close to 1, the prediction of mechanism of the LBD system needs to be improved because random selection of the terms gives almost the same performance.

3 Current Evaluation Approaches

Evaluating the performance of LBD systems is a fundamentally challenging task because if these systems are successful, by definition, they are capturing new knowledge that has yet to be proven useful. After a detailed analysis of the existing literature on LBD systems, we identified the following four different approaches used to evaluate LBD systems; replicating Swanson's discoveries, using statistical evaluation approaches, incorporating expert knowledge, and publishing in the medical domain. In this section, we will explain each evaluation approach in detail and discuss their advantages and disadvantages.

3.1 Replicating Swanson's Experiments

Even though the LBD systems are designed to produce new knowledge, measuring their performance by replicating the historical discoveries has been seen an effective evaluation approach by many LBD researchers. Swanson and Smalheiser published several different hypotheses about causally connected medical terms in the biomedical domain including *Migraine–Magnesium* [3], *Raynaud's Disease–Fish Oil* [4], *Alzheimer's Disease–Estrogen* [5], *Alzheimer's Disease–Indomethacin* [6], *Somatomedin C–Arginine* [7], and *Schizophrenia–Calcium Independent Phospholipase* A_2 [8]. Their discoveries have become gold standards for evaluation, and LBD researchers have measured the performance of their discovery systems by replicating Swanson's discoveries using the literature published before the original discovery dates. They have run their systems with Swanson's starting terms on the literature published prior to the discovery dates and reported overall success if one of the correlations generated by their systems matched Swanson's discovery.

Several researchers have used this strategy to evaluate the linking terms generated by their systems. Lindsay and Gordon [9] developed a process that followed the Swanson's discovery approach. They evaluated the performance of their process, in terms of precision and recall, for generating the linking terms, where Swanson's identified linking terms for *Migraine–Magnesium* example served as the gold standard. Gordon and Dumais applied latent semantic indexing to Swanson's discovery process [10]. They demonstrated the performance of their approach by replicating Swanson's *Raynaud's Disease* and *Fish Oil* discovery. Blake and Pratt applied a knowledge-based approach to identify and prune potential linking terms [11]. They replicated Swanson's *Migraine–Magnesium* example to evaluate their approach. However, all of these researchers focused on evaluating the linking terms by using Swanson's linking terms as the gold standard, and none pursued or evaluated how easy it would be identify the novel target term (e.g., *magnesium*), which is the main goal of LBD systems.

Weeber et al. also based their work on Swanson's approach [12]. They evaluated their literature-based discovery tool DAD by simulating Swanson's *Raynaud's Disease–Fish Oil* and *Migraine–Magnesium* examples. Their system supported both open and closed discovery approaches. In the open discovery approach, DAD first identified the linking terms that are directly connected to the starting terms, *Raynaud's Disease* and *Migraine*, and then identified the target terms that are connected to the linking terms identified in the first step. They reported which of the Swanson's linking terms DAD could identify and the ranks of *Fish Oil* and *Magnesium* in the final lists of target terms. In the closed discovery approach, they analyzed the starting term literature and the target term literature separately and identified the overlapping terms. They compared those terms with Swanson's linking terms and reported the results.

The most extensive evaluation of this type was done by Srinivasan [13]. She developed a literature based discovery system called Manjal. As Weeber et al.'s system, Manjal supports both open and closed discovery approaches. To evaluate her system, Srinivasan successfully replicated five of Swanson's discoveries including *Raynaud's Disease–Fish Oil*, *Migraine–Magnesium*, *Alzheimer's Disease–Indomethacin*, *Somatomedin C–Arginine*, and *Schizophrenia–Calcium Independent Phospholipase A2*. For each discovery, she reported the rank of the desired target term in the list of target terms generated by Manjal with the open discovery approach. She also reported the ranks of the desired linking terms identified by Manjal with the closed discovery approach.

Most recently, Hu et al. developed a prototype system called Bio-SbKDS based on Swanson's discovery approach [14]. They replicated Swanson's *Migraine–Magnesium* and *Raynaud Disease–Fish Oil* discoveries for evaluation purposes. He used *Migraine* and *Raynaud's Disease* as starting terms. They reported which of Swanson's linking terms their system could identify as linking terms and the ranks of *Magnesium* and *Fish Oil* in the final lists of target terms generated by their system.

In previous research, we also replicated Swanson's *Migraine–Magnesium* discovery to evaluate the capabilities of our system LitLinker [15]. As other researchers, we compared our linking terms with Swanson's linking terms and reported the rank of *Magnesium* in the final list of target terms.

The main advantage of this type of evaluation is the ease of designing it. In his papers, Swanson described each of his discoveries in great detail. The researchers use the information provided in those papers as a guide in designing their evaluations. For each discovery, the publication date of the corresponding paper serves as the original discovery date and the list of medical terms he used as links between his starting term and target term serves as a linking term gold standard.

Although all the researchers mentioned in this section have successfully replicated Swanson's discoveries, this type of evaluation is not complete because it does not inform us about the quality of the rest of the target terms identified by their systems. Depending on the approaches used to select the correlated terms, a literature-based discovery system might return hundreds or even thousands of terms as the target terms for a given starting term. Evaluating the whole system on only one of those target terms does not guarantee that the rest of the target terms also provide information with similar quality. As with information retrieval systems, an LBD system that returns a single helpful target term in a sea of unhelpful target terms is unlikely to be useful.

Another disadvantage of this approach is that the researchers are limited in their evaluations to the small number of discoveries published by Swanson. His discoveries mostly focused on diseases and their potential new treatments. Nevertheless, LBD tools can be used for various other tasks, such as identifying novel protein–protein interactions. Because the researchers know exactly what they are seeking as the desired target and linking terms in this limited set of discoveries, they can tune the parameters of their systems to be able to identify those terms. Such an approach might result in systems that perform well for the specific example cases but not well for other cases.

In addition, comparing the performance of different systems is one of the main objectives of system evaluation. However, replicating Swanson's discoveries does not allow detailed comparisons between different LBD systems. This evaluation method allows the researchers to say a system *A* is better than another system *B* if *A* simulates a selected discovery but *B* does not. However, if both *A* and *B* successfully simulate the given discovery successfully, it becomes impossible to determine which system is superior to the other.

3.2 Using Statistical Evaluation Methods

To overcome the drawbacks of the previous approach, some researchers have applied statistical evaluation methods to measure the overall performance of literature-based discovery systems for multiple target terms. As an example, Hristovski et al. performed a statistical evaluation of their system, BITOLA [16]. The purpose of their evaluation was to see how many of the potential discoveries made by their system at a specified point in time become realized at a later time. To accomplish this goal, they ran their system for the starting term *Multiple Seclerosis* on the set of documents published between 1990 and 1995. They checked the existence of the proposed discoveries in the set of documents published between 1996 and 1999 and calculated precision and recall. They used a very limited portion of the medical literature and reported the performance statistics of their system without comparing it to those of other systems.

To evaluate our system LitLinker, we used a similar but more extensive approach than Hristovski et al.'s approach; this approach enabled us to evaluate all correlations LitLinker generated. In our evaluation, for a given starting term, we measured whether LitLinker leads us to new discoveries in the more recently published medical literature. To accomplish this goal, we divided MEDLINE into two sets: (1) a baseline set including only publications before a selected cut-off date, and (2) a test set including only publications between the cut-off date and another later date. We ran LitLinker on the baseline set and checked the generated connections in the test set.

As an evaluation example, in [17], we ran LitLinker for the starting terms; *Alzheimer Disease*, *Migraine*, and *Schizophrenia* on a baseline set, which included only documents published before January 1, 2004 (cut-off date). We limited the linking terms and the target terms to only those terms in a semantic group listed in Table 1 because the goal of our experiments was to find novel connections between the selected *diseases* and *chemicals, drugs, genes, or molecular sequences*. We checked the existence of target terms generated by LitLinker in the test set that was composed of articles published between January 1, 2004 and September 30, 2005 (21 months).

Table 1 Semantic groups selected for our experiments

Linking term selection	Target term selection
Chemicals and drugs	Chemicals and drugs
Disorders	Genes and molecular Sequence
Genes and molecular sequence	
Physiology	
Anatomy	

To calculate precision and recall, for each starting term, we first retrieved the terms that co-occurred with the starting term in the test set but did not co-occur with the starting term in the baseline set. Then, we filtered the retrieved list of terms by using the semantic groups that we used for target term selections to find the ones that were chemicals, drugs, genes, or molecular sequences. We assumed that the terms in the remaining list would be new potential disease to gene or disease to drug treatment discoveries and used them as the target term gold standard for our precision and recall calculations.

In our research, we used our evaluation approach to compare two different methods for identifying linking or target terms based on a starting term, Z-Score [17] and MIM [18]. To accomplish this task, we first implemented the methods within our LitLinker framework. In our experiments, for each method, we ran LitLinker for ten randomly selected disease names on a baseline set, which includes only documents published before January 1, 2004. We created a target term gold standard for each disease from the test set documents published between January 1, 2004 and July, 31, 2006 (31 months).

We calculated precision and recall of both methods for each disease and ran statistical significance tests to measure the significance of the performance differences. We also used precision-recall graphs to compare different correlation methods. To draw precision-recall graphs, we used the ranked list of target terms generated by the two methods. We examined these lists of target terms starting from the top and selected intervals to calculate precision and recall with the formulas (1) and (2). Because we had ten different starting terms, to combine the results from each experiment, we calculated the average precision and recall for each interval. We also compared the prediction performances of both methods with that of pure random prediction with hypergeometric distribution as described in Sect. 2.3.

The main advantages of this type of evaluation are that the evaluation is fully automated, can be repeated for multiple starting terms, and enables comparison among different systems. On the other hand, its main drawback is that the calculated precision for target terms is the lower bound. The target term gold standard only includes the new correlations that are published between the cut-off date and the date of the experiment. It cannot include the correlations that will appear in the future. As a result, some of the target terms identified by the LBD system might become legitimate discoveries in the future but are considered incorrect target terms now. Another disadvantage is that this approach only evaluates the target terms without providing any information about the linking terms.

3.3 Incorporating Expert Opinion

As an alternative to the previous approaches, some researchers incorporated medical expert knowledge to the evaluation process of their LBD systems. Weeber et al., used their discovery system to investigate new potential uses for drug *thalidomide* with Swanson's open discovery approach [19]. One of the researchers involved in this study was a medical researcher with a background on pharmacology and immunology. For the starting term *thalidomide*, their system generated a list of linking terms that were constrained to be immunologic factors. They manually selected the promising linking terms with the involvement of the medical researcher. For the selected linking terms, their system generated a list of target terms that were constrained to be disease or syndrome names. The medical researcher manually assessed each of the selected diseases. In the assessment process, they tried to find additional bibliographic and other evidence for the linking terms between the *thalidomide* and the diseases identified as target terms. To accomplish this goal, for each disease, they first extracted the list of linking terms that connect the disease to *thalidomide*. Next, they extracted the sentences that included *thalidomide* and the extracted linking terms and the sentences that included the linking terms and the disease. They provided those sentences to the medical expert for assessment. Based on the assessment, they compiled a list of four diseases; *chronic hepatitis C*, *myasthenia gravis*, *helicobacter pylori induced gastritis*, *acute pancreatitis* for which the researchers hypothesized that *thalidomide* could be an effective treatment.

Srinivasan and Libbus evaluated their system Manjal by using a semi-automated approach with experts. In their experiment, they used *turmeric*, a widely used spice in Asia, as their starting term. The aim of their experiment was to identify diseases where *turmeric* could be useful in the treating them. They ran Manjal for the starting term *turmeric*, and, with the selected thresholds, Manjal identified 26 terms as the linking terms, L_1. To evaluate the linking terms in L_1, a medical researcher identified a second set of linking terms, L_2, after reading the documents about *turmeric*. There were 27 terms in L_2. They used this manually created list as the linking term gold standard. They compared L_1 with L_2 and calculated recall and precision with the following formulas:

$$\text{Precision} : P = \frac{\|L_1 \cap L_2\|}{\|L_1\|}. \tag{7}$$

$$\text{Recall} : R = \frac{\|L_1 \cap L_2\|}{\|L_2\|}. \tag{8}$$

Manjal generated two sets of target terms; one from the automatically generated linking terms and one from the manually selected linking terms. They used the second set as the target term gold standard to evaluate the first set and reported precision and recall. In addition to reporting precision and recall, they did a detailed citation analysis and described the potential use of *turmeric* in the treatment of *retinal diseases*, *Crohn's disease*, and *spinal cord injuries*. In contrast to the statistical approach described in the previous section, the advantage of Srinivasan and Libbus's approach is that it allows us to evaluate the linking terms in addition to the target terms. However, the evaluation highly depends on the subjective decision of the medical researcher in deciding which terms are correlated with the starting term. This decision is crucial because it also directly effects the selection of the terms in the target term gold standard. It is also unclear whether the gold standard set of target terms reflects a true gold standard because no checking has been done on those target terms.

Wren et al. also incorporated medical expert knowledge into the evaluation process [20]. The researchers who contributed to this study had a medical background. They ran their literature-based discovery approach for the starting term *cardiac hypertrophy* and identified a total of 2,102 linking terms and 19,718 target terms. To evaluate their approach, they performed laboratory tests for the third ranked target term, *chlorpromazine*. *Chlorpromazine* is a chemical that is used as an anti-psychotic and anti-emetic drug. In their lab experiments, they looked for an association between *chlorpromazine* and *cardiac hypertrophy*. They gave 20mg/kg/day per day *isoproterenol* by osmotic minipump to two groups of mice, with one group additionally receiving 10mg/kg/day per day *chlorpromazine*. Their results showed that the amount of *cardiac hypertrophy* was significantly reduced in the *isoproterenol* plus *chlorpromazine* treated mice in comparison to the control group only given *isoproterenol*. They reported that *chlorpromazine* could reduce *cardiac hypertrophy* by showing their experimental results with mice as evidence. Their work is an excellent example of how literature-based discovery tools can be integrated to medical researcher's real-life research activities.

The main advantage of this type of evaluation is the involvement of the medical researchers, who are the real users of the LBD systems into the evaluation process. To identify what medical researchers find interesting or not interesting could inform LBD system designers while they upgrade the algorithms or the other approaches they use in the discovery process. The downside is the high cost of evaluation. Weeber et al. reported that their manual effort while evaluating the output of their system consisted of several one hour sessions during a two week period. Such an evaluation is also hard to quantify, and thus hard to use to compare different LBD systems. Because the aim of LBD tools is to identify novel correlations, disagreements on the interestingness of the correlations could arise if multiple medical researchers are involved in the evaluation process.

3.4 Publishing in the Medical Domain

Another approach that is used to evaluate LBD systems is publishing the discoveries in medical journals or presenting them in the medical domain. This evaluation approach is a very powerful yet a very challenging one. Publishing in the medical domain requires the flexibility to write for the medical audience, but the overall benefit is clear: validation of work, impact on the science, external visibility for LBD research, and the chance to gain new collaborators. This type of evaluation is not commonly used in LBD research. Among all LBD researchers, Swanson is the only researcher who could publish his discoveries in the medical journals. In addition to Swanson's personal interest in medicine, his close collaboration with Smalheiser who is a medical doctor and neuroscientist, resulted in various publications [3–8, 21].

4 User Interface Evaluation

The success of an LBD system in facilitating new discoveries depends on its interface's ability to inform and engage its users as they attempt to interpret and evaluate the proposed connections. The amount of data produced by an LBD system is usually immense. As an example, when LitLinker replicated Swanson's *Migraine–Magnesium* discovery, it processed over 4 million documents. It generated 349 linking terms and 545 target terms with 57,622 possible starting term-linking term and linking term-target term combinations. To be able to handle the amount and complexity of the output data, one of the primary objectives of an LBD system interface must be to promote user comprehension of numerous complex relationships among the terms involved in each proposed connection in an effective way. The interface must also provide flexible navigation and a level of detail appropriate to the scope of each view without obscuring data necessarily for evaluation purposes. And most importantly, the interface should help researchers incorporate the LBD system's results into their own research discovery process. To accomplish those objectives requires the involvement of real users into the interface design process. One way to involve users is by conducting usability evaluations and changing the interface design according to the feedback collected from the participants of the evaluation.

We designed a web-based graphical interface for LitLinker[1]. Our aim in developing an interface was to allow researchers to carefully assess the potential connections generated by LitLinker. We first developed a prototype interface and conducted a usability evaluation with ten participants, including nine graduate students and one faculty member [22]. The evaluation consisted of three parts: a general introduction, a task-based questionnaire, and an interview. The participants used LitLinker with *Migraine* as the starting term, to complete a task-based questionnaire. The tasks were designed to evaluate each participant's ability to find specific data, to navigate the interface, and to compare the strengths of connections. Participants were asked to talk aloud and as they completed the tasks. The interviewer observed without answering questions and noted any difficulties the participants experienced. After participants completed the questionnaire, we interviewed them to discover aspects of the interface that were confusing or were particularly helpful. We identified many design problems during this usability evaluation and modified our interface to increase its usability.

Similarly, Smalheiser et al. evaluated their LBD system, Arrowsmith as part of a 5 year neuroscience project at University of Illinois – Chicago [23]. The goal

[1] Available at: http://litlinker.ischool.washington.edu/index.jsp

of their evaluation study included making scientific discoveries, publishing papers, and identifying new research directions. In contrast to our study, they did not recruit human subjects or study their behavior on standardized tasks. Rather, the medical researchers who participated in the study chose the search topics and observed the outcomes. Each participant was given an electronic notebook to record opportunities for conducting Arrowsmith searches, whether they arose from laboratory experiments, from attending conferences, or from discussions with other researchers, and to record the details of completed Arrowsmith searches. Participants sent the notebook entries via e-mail to the researchers and the researchers called the participants every week to monitor the course of their scientific work, to learn more about the completed searches, to receive suggestions for improving the interface, and to document the follow-up of completed searches. Based on the input they received from the participants, they updated the Arrowsmith interface. They also focused on information seeking needs and strategies of medical researchers as they formulate new hypotheses.

5 Conclusion

LBD systems have great promise for improving medical researchers' efficiency while they seek information in the vast amount of literature available to them. Although many online LBD systems are available, they are not in routine use. For a wider usage of LBD systems, effective evaluation is essential. Evaluation will not only help to identify which algorithmic approaches work best for LBD, but also provide information about how discovery systems can best enhance the real-life work processes of medical researchers. In this chapter, we summarized the current evaluation approaches used to evaluate LBD systems and their interfaces, but more research on evaluation methods that standardize system comparisons and explore user behavior is needed.

Acknowledgements The National Science Foundation, award #IIS-0133973, funded this work.

References

1. Baeza-Yates, R. and Ribeiro-Neto, B., *Modern Information Retrieval.* 1999: ACM Press, Addison-Wesley
2. Bradley, A.P., *The Use of the Area Under the Roc Curve in the Evaluation of Machine Learning Algorithms.* Pattern Recognit., 1997. **30**(7): 1145–1159
3. Swanson, D.R., *Migraine and Magnesium: Eleven Neglected Connections.* Perspect. Biol. Med., 1988. **31**: 526–557
4. Swanson, D.R., *Fish Oil, Raynaud's Syndrome, and Undiscovered Public Knowledge.* Perspect. Biol. Med., 1986. **30**(1): 7–15
5. Smalheiser, N.R. and Swanson, D., *Linking Estrogen to Alzheimer's Disease: An Informatics Approach.* Neurology, 1996. **47**(3): 809–810

6. Smalheiser, N.R. and Swanson, D., *Indomethacin and Alzheimer's Disease*. Neurology, 1996. **46**(2): 583
7. Swanson, D.R., *Somatomedin C and Arginine: Implicit Connections Between Mutually Isolated Literatures*. Perspect. Biol. Med., 1990. **33**(2): 157–186
8. Smalheiser, N.R. and Swanson, D., *Calcium-Independent Phospholipase A2 and Schizophrenia*. Arch. Gen. Psychiatry, 1998. **55**: 752–753
9. Lindsay, R.K. and Gordon, M.D., *Literature Based Discovery by Lexical Statistics*. J. Am. Soc. Inf. Sci., 1999. **49**(8): 674–685
10. Gordon, M.D. and Dumais, S., *Using Latent Semantic Indexing for Literature Based Discovery*. J. Am. Soc. Inf. Sci., 1998. **49**(8): 674–685
11. Blake, C. and Pratt, W., Automatically Identifying Candidate Treatments from Existing Medical Literature. In *Proceedings of AAAI Spring Symposium on Mining Answers from Texts and Knowledge Bases*. 2002. Stanford, CA
12. Weeber, M., Klein, H., and de Jong-van den Berg, L.T.W., *Using Concepts in Literature Based Discovery: Simulating Swanson's Raynaud–Fish Oil and Migraine–Magnesium Examples*. J. Am. Soc. Inf. Sci., 2001. **52**(7): 548–557
13. Srinivasan, P., *Generating Hypotheses from MEDLINE*. J. Am. Soc. Inf. Sci., 2004. **55**(5): 396–413
14. Hu, X., Li, G., Yoo, I., Zhang, X., and Xu, X., A Semantic-Based Approach for Mining Undiscovered Public Knowledge from Biomedical Knowledge. In *Proceedings of IEEE International Conference on Granular Computing*. 2005. Beijing
15. Pratt, W. and Yetisgen-Yildiz, M. L., Capturing Connections Across the Biomedical Literature. In *Proceedings of International Conference on Knowledge Capture (K-Cap'03)*. 2003. Florida
16. Hristovski, D., Stare, J., Peterlin, B., and Dzeroski, S., Supporting Discovery in Medicine by Association Rule Mining in Medline and UMLS. In *Proceedings of Medinfo Conference*. 2001. London, England
17. Yetisgen-Yildiz, M. and Pratt, W., *Using Statistical and Knowledge-Based Approaches for Literature Based Discovery*. J. Biomed. Inform., 2006. **39**(6): 600–611
18. Wren, J.D., *Extending the Mutual Information Measure to Rank Inferred Literature Relationship*. BMC Bioinformatics, 2004. **5**(1): 145
19. Weeber, M., Vos, R., Klein, H., de Jong-van den Berg, L.T.W., and Aronson, A.R., *Generating Hypotheses by Discovering Implicit Associations in the Literature: A Case Report of a Search for New Potential Therapeutic Uses for Thalidomide*. J. Am. Med. Inform. Assoc., 2003. **10**(3): 252–259
20. Wren, J.D., Bekeredjian, R., Stewart, J.A., Shohet, R.V., and Garner, H.R., *Knowledge Discovery by Automated Identification and Ranking of Implicit Relationships*. Bioinformatics, 2004. **20**(3): 389–398
21. Swanson, D.R., *Atrial Fibrillation in Athletes: Implicit Literature-Based Connections Suggest that Overtraining and Subsequent Inflammation may be a Contributory Mechanism*. Med. Hypotheses, 2006. **66**(6): 1085–1092
22. Skeels, M.M., Henning, K., Yetisgen-Yildiz, M., and Pratt, W., Interaction Design for Literature-Based Discovery. In *Proceedings of the International Conference for Human-Computer Interaction (CHI'05)*. 2005. Portland, WA
23. Smalheiser, N.R., et al., *Collaborative Development of the Arrowsmith Two Node Search Interface Designed for Laboratory Investigators*. J. Biomed. Discov. Collab., 2006. **1**(8)

Factor Analytic Approach to Transitive Text Mining using Medline Descriptors

J. Stegmann and G. Grohmann

Abstract Matrix decomposition methods were applied to examples of non-interactive literature sets sharing implicit relations. Document-by-term matrices were created from downloaded PubMed literature sets, the terms being the Medical Subject Headings (MeSH descriptors) assigned to the documents. The loadings of the factors derived from singular value or eigenvalue matrix decomposition were sorted according to absolute values and subsequently inspected for positions of terms relevant to the discovery of hidden connections. It was found that only a small number of factors had to be screened to find key terms in close neighbourhood, being separated by a small number of terms only.

It is concluded that in literature-based discovery processes the decomposition methods combined with human inspection of the created factors may play an important role provided MeSH descriptors are analysed.

Keywords: Text mining · Swanson linking · Hypothesis generation · Matrix decomposition · Medline descriptors

1 Introduction

Transitive text mining tries to establish meaningful links between the main concepts of non-overlapping literature sets. The basic notion of this kind of literatures as 'Complementary But Disjoint' (CBD) and several examples of literature-based

---

J. Stegmann
Buergipfad 24, 12209 Berlin, Germany
johannes.stegmann@onlinehome.de

G. Grohmann
Charité - Institute of Medical Informatics, 12203 Berlin, Germany
guenter.grohmann@charite.de

P. Bruza and M. Weeber (eds.), *Literature-based Discovery*,
Springer Series in Information Science and Knowledge Management 15.

'discoveries' from the medical literature have been published by Don Swanson [1–5]. Hence, this particular type of text mining has also been named 'Swanson Linking' (SL) [6].

Basically, SL deals with three different types of literature: (a) a problem-based literature – e.g. describing a disease – is referred to as "source"; (b) a literature not being mentioned in the source literature but possibly contributing to problem solving is called "target"; (c) a literature representing a major concept which is relevant to and occurs in both, source and target literature, is labeled "intermediary" [7]. The discovery process might normally proceed from source to target via intermediary; however, the reverse order is naturally conceivable, and any coherent literature set regarded as "intermediary" may be explored for source and target concepts simultaneously [8]. Classical examples of CBD literatures are *Raynaud's Disease – Fish Oil* with *Blood Viscosity* as intermediary [1], and *Migraine – Magnesium* with *Spreading Depression* or *Epilepsy* as intermediaries [2]. Other examples include the literature pairs *Somatomedin C – Arginine* [5], *Alzheimer's Disease – Indomethacin, Estrogen* [9, 10], and *Schizophrenia – Phospholipase* A_2 [11].

Different methods have been applied to detect possibly useful links between 'non-interactive medical literatures' [3], subjecting to various statistical procedures either the words and phrases taken from document titles [7] and from titles and abstracts [12–14], or the descriptors assigned to the indexed documents [6, 15–17]. All types of approaches try to find intermediary terms and concepts for retrieval of the respective literature which contains the otherwise non-interactive 'source' and 'target' terms. The natural language-based methods attempt to find interesting terms/concepts on top of ranking lists, with [14] or without [7, 12, 13] application of the UMLS (Unified Medical Language System) metathesaurus [18] as a semantic filter. The descriptor-based approaches extract the Medical Subject Headings (MeSH [19]) assigned to each document indexed in Medline from the downloaded documents and try to find intermediary and target terms either in two-dimensional cluster representations of the MeSH descriptors [6, 16, 17] or on ranked lists after application of semantic UMLS filters [15] (see also [20]).

The different types of transitive text mining described so far can be classified as 'bottom-up' approaches where an initial knowledge space spanned by one or a few source word(s) is explored to find intermediary literature which in turn spans the term space harboring both, source and possible target concepts. To a second class of attempts to attain literature-based discoveries (in the sense of building hypotheses) belong recent large-scale efforts: one uses all electronically available Medline records to construct a network of biomedically relevant objects (obtained by matching the document texts against a pre-compiled composite dictionary of, e.g., gene and disease names) and to find promising implicit links by shared relationships between not directly connected objects [21]. Other projects use the complete thesaurus (or significant parts of it) of MeSH descriptors and precompile their mutual relationships using the whole Medline database and the UMLS for application of semantic filters [22], or add gene dictionaries for prediction of implicit disease–gene relationships [23, 24].

In our previous work we were able to show that SL can be successfully performed by co-occurrence analysis of the MeSH descriptors assigned to the documents of the investigated PubMed literature sets [6, 16, 17]. We applicated the cluster algorithm described by Callon et al. [25] to the MeSH terms extracted from source and intermediary literature sets of the known pairs *Raynaud's Disease – Fish Oil*, *Migraine – Magnesium* [6, 16], and *Schizophrenia – Phospholipase* A_2 [17]. Displaying the MeSH term clusters in two-dimensional diagrams according to internal and external link strength (density and centrality, respectively) we found that in some cases relevant source, intermediary and target terms are located in clusters with salient positional and/or numerical characteristics. For example, in the diagram derived from the intermediary *Blood Viscosity* literature set we see the two clusters harboring the source (MeSH) term *Raynaud Disease* and the target (MeSH) terms *Fish Oil* and *Eicosapentaenoic Acid* in close neighbourhood; both clusters also have approximately the same centrality/density ratio.

Having performed a similar analysis of the pre-1996 literature on prions [6, 16] we were able to deduce from the resulting cluster diagrams several hypotheses being discussed in the more recent prion research literature.

The MeSH term cluster analysis is also a core feature of our recently released *Charité MLink* [26] text mining tool [27].

However, not in all cases so far investigated by us are the "interesting" terms (descriptors which pave the way from source to target) found in clusters with specific characteristics, and high numbers of terms derived from large literature sets consisting of thousands of documents hamper easy cluster localisation and screening. Therefore, we are interested in alternative methods which may place the relevant terms in close vicinity to each other. A promising approach is to reduce the dimensionality of the semantic space spanned by documents and terms. Gordon and Dumais [13] applied singular value decomposition (svd) – introduced as a tool for 'Latent Semantic Analysis' (LSA) by Deerwester et al. [28] – to establish the chain from *Raynaud's Disease* to *Fish Oil*. These authors analysed words and phrases taken from titles and abstracts of the respective literature sets. After decomposition of the rectangular term-by-document matrices using svd they re-built them on a considerably lower dimensional level and subsequently determined the cosine-based similarity between terms. According to that measure they found intermediary concepts near to the source term 'Raynaud' whereas known target terms like 'Eicosapentaenoic Acid' and 'Fish Oil' were still located far apart from the source term, thus disabling a quick successful inspection of the term lists. Yet, the capability of LSA to uncover implicit relationships was demonstrated analysing genes on the basis of their textual representations as provided by Medline abstracts [29].

We tested the decomposition technique on the semantic space spanned by the MeSH terms assigned to the documents of the literature sets under investigation. We analysed the factors (i.e. the eigenvectors created by matrix decomposition) directly by manual inspection, omitting the further steps of Gordon and Dumais [13]. We show in this communication that relatively easy and fast screening of only the first few factors finds relevant intermediary and target terms in close proximity to source terms as determined by the absolute values of the corresponding factor loadings.

As examples, we first analyse literature sets around two classical examples: (a) *Raynaud's Disease* [1], with *Blood Viscosity* and *Platelet Aggregation* as intermediary terms, *Eicosapentaenoic Acid*, *Fish Oils*, *Arginine* and *Nitric Oxide* [6] as target terms; (b) *Migraine* [2], with *Spreading Cortical Depression* and *Epilepsy* as intermediaries and *Magnesium* as target. Next, we examine (c) the example of non-interactive literatures recently described by Wren et al. [21] who – from large-scale network analysis – found hints to implict relationships between *Cardiac Hypertrophy* and the anti-psychotic drug *Chlorpromazine*. From a list of shared relationships between both concepts (published by these authors as online supplemental material [30]) we chose several terms and found that they act successfully as intermediaries in our factor analysis. Finally, we (d) explore literature sets spotting *Multiple Sclerosis* and *Erythropoietin* as CBD partners, the latter being currently discussed as a potential neuroprotective agent (e.g. [31, 32]). Using the matrix decomposition method described in the present communciation we identified *Nitric Oxide Synthase* as an intermediary term and literature.

2 Methods

2.1 Data

Document sets were retrieved from PubMed [33] performing appropriate title searches. Retrieval details are given in Sect. 3. Documents were downloaded in PubMed's MEDLINE format. Descriptors were extracted by means of homemade Perl scripts (run by perl version 5.004 under IRIX 6.5.28f) from both, the MH and RN fields of the documents as described [6] (we use "MeSH" somewhat loosely: it comprises MH and RN terms). Those named entities of the RN fields which are different from MH descriptors contribute to an extent of well below 10% (mostly below 5%) to the list of distinct descriptors of a set (data not shown). Subheadings and multiple occurrences of descriptors in a document were ignored. Descriptors occurring in one document only were omitted.

2.2 Matrix Decomposition

A binary document-by-term matrix was produced for each document set with record numbers as rows, descriptors as columns, and "0" or "1" as values of the matrix cells indicating whether the respective descriptor is or is not contained in the corresponding record. These matrices were subjected to singular value or – after appropriate modification – to eigenvalue decomposition (svd or evd, respectively), using the software package R [34] (version 2.2.0, run under MSWinXP). Please, note that we always started with a document-by-term matrix (see above) in contrast to Gordon and Dumais [13] who analysed term-by-document matrices. Application of svd to a document-by-term matrix X of the type described above results in three matrices U, D, V (following the notation of the software package), where D is a diagonal matrix with the singular values. U and V are orthogonal matrices where U contains as rows the record numbers and as columns the left singular vectors (the factors constituted by the documents), and V contains as rows the descriptors and as columns the right singular vectors (the factors constituted by the descriptors). The matrix elements of U and V are the respective factor loadings. The three matrices (D, U, V) are necessary to rebuild the original matrix on a lower dimensionality level. In preliminary experiments, we tested the *Blood Viscosity* literature set whether a reduction in dimensionality resulted in a salient similarity of the corresponding source and target descriptors (*Raynaud Disease – Fish Oils, Eicosapentaenoic Acid*) but failed (data not shown). Therefore, we analysed matrix V directly which can also be created by eigenvalue decomposition (evd) of the square symmetric matrix $X^T X$. Thus, in all cases shown we left-multiplied the original document-by-term matrix X with its transposed X^T and obtained the symmetric matrix Y with the co-occurrence frequencies of the descriptors. Y was then subjected to evd to produce V. The resulting eigenvectors are identical to the singular vectors of svd-created V, and the resulting eigenvalues (measures of the variance extracted from the data by the respective factors) are the squared singular values [28]. By convention, the singular values and eigenvalues are sorted in decreasing order, and the first (most-left) singular (eigen-) vector corresponds to the highest singular (eigen-) value.

2.3 Factor Screening

Prior to visual inspection the elements of matrix V were converted to absolute values (indicating the extent to which the corresponding descriptors contribute to the respective factors, see [35]) and sorted column by column in decreasing order. Beginning with the first factor inspection started at the position of the "guiding term(s)". In the present investigation guiding terms are "source terms", i.e. the MeSH descriptors designating the disease being the "source" of the discovery process (as *Raynaud's Disease*, *Migraine*, etc.). The factor elements above and below (i.e., "in the neighbourhood" of) the source descriptors were screened for intermediary or target terms (analysing factors derived from source or intermediary literature sets, respectively).

3 Results

3.1 Raynaud's Disease

Figure 1 illustrates the factor screening with the appropriate part of factor 10 of the eigenvector matrix derived from the *Blood Viscosity* intermediary literature set. The eigenvector elements have been converted to absolute values and the respective d = 9

d = 28

Pos.	MeSH descriptors	Factor loadings (absolute values)
60	Hemodilution	0 . 0357994841822661
61	Arteries	0 . 0348064869815320
62	Colloids	0 . 0339972747368806
63	Methods	0 . 0318994228076309
64	Raynaud Disease	0 . 0318162159389412
65	Diabetic Angiopathies	0 . 0313996783916599
66	Insulin	0 . 0310927490267344
67	Perfusion	0 . 0309766748671588
68	Plasma	0 . 0304961101716611
69	Intermittent claudication	0 . 0304187422028703
70	Snakes	0 . 0297750960966340
71	Venoms	0 . 0297750960966340
72	Iron	0 . 0294382154034097
73	Fatty Acids, Unsaturated	0 . 0291045562973698
74	Eicosapentaenoic Acid	0 . 0291045562973698
75	Serum Albumin	0 . 0285885038109841
76	Glucose	0 . 0273433766104969
77	Rats, Inbred strains	0 . 0270774228087507
78	Creatinine	0 . 0268549399732287
79	Sodium	0 . 0268113400323863
80	Oxygen Consumption	0 . 0264810443785870
81	Cardiovascular Diseases	0 . 0263689507529746
82	Random Allocation	0 . 0246847696705744
83	Pottassium	0 . 0246614766785590
84	Iliac Artery	0 . 0246216496147969
85	Rotation	0 . 0238328388320026
86	Gelatin	0 . 0237639049296123
87	Oxygen	0 . 0234868334744486
88	Plasma substitutes	0 . 0234070186919323
89	Anticoagulants	0 . 0232874480931843
90	Ear	0 . 0230951722588649
91	Drug Combinations	0 . 0226196952348161
92	Coronary Disease	0 . 0226060239195273
93	Fish Oils	0 . 0224276641261152
94	Arteriosclerosis	0 . 0224084131039026

Fig. 1 Illustration of factor screening.
The appropriate part of factor 10 of the eigenvector matrix of the *Blood Viscosity* intermediary literature set is displayed after sorting of the MeSH descriptors according to descending order of the absolute values of the factor loadings.
Screening starts at the position of the source term (*Raynaud Disease*).
Positions of target terms (*Eicosapentaenoic Acid* and *Fish Oils*) are indicated.
d (distance) is the number of descriptors between source and target term.
Methodical details: see Sect. 2.
Retrieval details: see Table 2

descriptors have been sorted in decreasing order of the values (see Sect. 2). Screening starts at the position of the guiding term, i.e. the source term *Raynaud Disease* which is found on position 64. The target term *Eicosapentaenoic Acid* is on position 74. The distance from *Raynaud Disease* to *Eicosapentaenoic Acid* is 9 (nine terms are in between). The distance from the source term to the other relevant target term *Fish Oils* which is on position 93 is 28. Thus, *Eicosapentaenoic Acid* would certainly attract the attention of a surveyor who performs a screen of factor 10 involving no more than 20 terms (roughly 20% of the total number of descriptors, see Table 2) around the source term, but not *Fish Oils* which is in this factor too far away from *Raynaud Disease*.

A factor analysis of the source literature set on *Raynaud's Disease* is shown in Table 1 which lists the positions of the guiding (source) term *Raynaud Disease* and the known relevant intermediary descriptors *Blood Viscosity* and *Platelet Aggregation* within the first 20 factors after sorting of the descriptors in descending order according to the absolute values of their contributon to the respective factor. Black table cells indicate distances of up to ten terms between source term and intermediary descriptor, grey table cells indicate source-intermediary distances of 11–20

Table 1 Factor analysis of the *Raynaud's Disease* source literature set.
Retrieval details: PubMed title search for **raynaud***, limited to publication years 1966–1985.
Retrieval date: June 21, 2004.
Descriptor extraction, eigenvalue decompositon, factor screening: see Sect. 2.
Size of the document-by-term matrix: 801×464.
Source term: *Raynaud Disease*.
Intermediary terms: *Blood Viscosity* (BV), *platelet aggregation* (PA).
Distance: number of terms between source and intermediary term (*black cells*: 0–10, *grey cells*: 11–20)

Factor	Position of source term	Position of intermediary term		Distance	
		BV	PA	BV	PA
1	1	34	145	32	143
2	1	20	280	**18**	278
3	10	37	411	26	400
4	94	135	211	40	126
5	53	56	174	**2**	120
6	26	323	376	205	349
7	383	51	113	331	269
8	9	29	253	**19**	243
9	138	57	117	80	**20**
10	73	121	61	47	**11**
11	16	35	73	**18**	56
12	42	103	263	60	220
13	161	68	375	92	213
14	64	36	448	27	383
15	55	25	130	29	74
16	62	18	327	43	264
17	167	318	110	150	56
18	56	260	335	203	278
19	215	5	214	209	**0**
20	222	38	177	183	44

Table 2 Factor analysis of the *Blood Viscosity* intermediary literature set.
Retrieval details: PubMed title search for **"blood viscosity"**, limited to publication years 1966–1985.
Retrieval date: June 21, 2004.
Size of the document-by-term matrix: 502 × 393.
Source term: *Raynaud Disease.*
Target term: *Eicosapentaenoic Acid.*
Other details: see Table 1

Factor	Position of source term	Position of target term	Distance
10	64	74	9
12	54	64	9
23	35	37	1
25	92	108	15
27	34	45	10
29	72	78	5

terms. Within the 20 factors we find seven distances of 20 or less and two distances of 10 or less. It seems reasonable that screening of the first 20 factors is sufficient to detect the important intermediary descriptors *Blood Viscosity* and/or *Platelet Aggregation*. We think that 5% is an upper limit for screening of factors and terms; thus, it seems reasonable to limit the survey of the factor matrix constituted by the 464 MeSH descriptors of the *Raynaud's Disease* literature set to 20 terms above and below the source term and to 20 factors. We also think that 30–40 terms/factors may be an upper absolute limit for a short and quick look up of a factor matrix.

Results from factor analyses of the intermediary literature sets on *Blood Viscosity* and *Platelet Aggregation* are shown in Tables 2 and 3, respectively. Factors with relatively short distances between the source term *Raynaud Disease* and the target terms *Eicosapentaenoic Acid* (Tables 2 and 3), *Fish Oils*, *Arginine*, *Nitric Oxide* (Table 3) are listed, and the corresponding term positions are indicated (*Arginine* and *Nitric Oxide* have been identified earlier as potential target terms [6]). Here (Tables 2 and 3), the "yield" of short distances between source and target terms within the limit of 20 factors is considerably lower than in the case of the source literature (see Table 1). We cannot deduce any regularities from the number of documents (*Blood Viscosity*: 502, *Platelet Aggregation*: 2,636) or MeSH descriptors (*Blood Viscosity*: 393, *Platelet Aggregation*: 1,532) (Tables 2 and 3) whereas *Raynaud's Disease* exhibiting a higher number of short distances between the relevant terms comprises 801 documents and 464 MeSH descriptors (Table 1). A role may play the relative term frequency, i.e. the fraction of documents containing the relevant terms. Four percent of the source literature (*Raynaud's Disease*) documents contain the descriptor *Blood Viscosity*. However, the source and target terms occur in less than 0.5% of the intermediary *Platelet Aggregation* documents. In less than 1% of the

Table 3 Factor analysis of the *Platelet Aggregation* intermediary literature set.
Retrieval details: PubMed title search for **"platelet aggregation"**, limited to publication years 1966–1985.
Retrieval date: November 11, 2004.
Size of the document-by-term matrix: $2,636 \times 1,532$.
Source term: *Raynaud Disease*.
Target terms: *Eicosapentaenoic Acid* (EPA), *Fish Oils* (FO), *Arginine* (Arg), *Nitric Oxide* (NO).
Other details: see Table 1

Factor	Position of source term	Position of target term	Distance
12	447	471 (EPA)	23
20	762	790 (FO)	27
21	775	754 (Arg)	**20**
27	964	985 (FO)	**15**
31	526	524 (Arg)	**1**
32	776	787 (FO)	**10**
37	614	617 (NO)	**2**

Blood Viscosity literature documents occur the target terms (*Eicosapentaenoic Acid*, *Fish Oils*) whereas the source term *Raynaud Disease* has a relative frequency of 3.2% (data not shown).

On the other hand, within the first 30 factors derived from the *Blood Viscosity* literature set we see five times very short distances of no more than ten terms between source and target term (Table 2), and several of the first 40 factors (only 2.6% of total) of the *Platelet Aggregation* factor matrix show source and target terms in close or very close vicinity (Table 3). Thus, the results obtained from the eigenvalue decomposition technique applied to source and intermediary literature sets of the first classical SL example *Raynaud's Disease* [1] encouraged us to use the method for analysis of other SL literatures.

3.2 *Migraine*

The screening of the factors produced by evd of the *Migraine* source literature starts at the position of the source descriptor *Migraine*. Look up is for the intermediary terms *Spreading Cortical Depression* and *Epilepsy*. Table 4 lists those of the first 30 factors in which source and intermediary terms are no more than 20 terms apart. In several factors the source and one of the intermediary terms are very close together; in factor 16 even both intermediary terms are less than ten terms apart from the source descriptor (Table 4). Analyses of the two intermediary literatures (*Spreading Cortical Depression*, *Epilepsy*) are shown in Tables 5 and 6. Whereas we find four times very short distances (less than 10 terms apart) between the source term

Table 4 Factor analysis of the *Migraine* source literature set.
Retrieval details: PubMed title search for **migraine**, limited to publication years 1966–1987.
Retrieval date: November 10, 2004.
Size of the document-by-term matrix: $2,583 \times 1,090$.
Source term: *Migraine*.
Intermediary terms: *Epilepsy* (Epi), *Spreading Cortical Depression* (SCD).
Other details: see Table 1

Factor	Position of source term	Position of intermediary term	Distance
3	25	13 (Epi)	**11**
4	155	162 (SCD)	**6**
6	38	58 (Epi)	**19**
7	36	47 (Epi)	**10**
13	62	64 (SCD)	**1**
14	42	26 (SCD)	**15**
16	45	36 (Epi)	**8**
		38 (SCD)	**6**
17	37	21 (Epi)	**15**
22	110	130 (Epi)	**19**

Table 5 Factor analysis of the *Spreading Cortical Depression* intermediary literature set.
Retrieval details: PubMed title search for **"spreading cortical depression" OR "spreading depression"**, limited to publication years 1966–1987.
Retrieval date: November 10, 2004.
Size of the document-by-term matrix: 326×293.
Source term: *Migraine*.
Target term: *Magnesium*.
Other details: see Table 1

Factor	Position of source term	Position of target term	Distance
1	67	60	**6**
2	39	26	**12**
5	21	19	**1**
6	15	24	**8**
11	82	74	**7**

Migraine and the target term *Magnesium* within the first 11 factors of the *Spreading Cortical Depression* factor matrix (Table 5), we do not find reasonable distances (less than 30 terms) between source and target in the first 25 factors of the *Epilepsy* factor matrix (Table 6). Factor 26 and 29, however, contains source and target rather close together (2 and 12 terms in between). As mentioned above, the relative term

Table 6 Factor analysis of the *Epilepsy* intermediary literature set.
Retrieval details: PubMed title search for **epilepsy**, limited to publication years 1966–1987.
Retrieval date: November 12, 2004.
Size of the document-by-term matrix: $6,682 \times 2,154$.
Source term: *Migraine*.
Target term: *Magnesium*.
Other details: see Table 1

Factor	Position of source term	Position of target term	Distance
26	709	706	2
29	273	260	12

frequency may contribute to the number of short distances between the terms in question: both, *Migraine* and *Magnesium* occur in 3.4% of the *Spreading Cortical Depression* documents but only in 0.7% (*Migraine*) and 0.1% (*Magnesium*) of the *Epilepsy* documents. A 2.7% of the *Migraine* documents contain the intermediary term *Epilepsy* (exhibiting several short distances to the *Migraine* term in the first factors, see Table 4). However, *Spreading Cortical Depression* is represented by only 0.8% of the *Migraine* documents but has also several short distances to the source term (see Table 4). Thus, other factors (e.g. number of distinct descriptors per document) may influence the distance patterns.

3.3 Cardiac Hypertrophy

The concept of *Cardiac Hypertrophy* is represented by several MeSH descriptors. Short distances (up to 20 terms apart) of these terms to some intermediary items (taken from [30] and being identical with MeSH descriptors) appearing within the first 20 factors are listed in Table 7. Relative term frequencies range from 0.9% for *Adenosine Triphosphate* to 2.3% for *Verapamil*. Short or very short distances are apparent in at least three factors for each intermediary term except *Protein Kinases* which has one short distance (Table 7). When the first 30 factors are analysed the descriptor *Protein Kinases* exhibits three times short or very short distances, and each of the other intermediary terms is close or very close to a source descriptor in at least seven factors (not shown). From the results seen so far it can be concluded that the method is sensitive for small and large literature sets as well as low and high(er) relative descriptor frequencies ranging from 0.1 to 4%.

Table 8 displays the instances of short distances (40 terms or less) between the target descriptor *Chlorpromazine* and the *Cardiac Hypertrophy* source descriptor(s) within the first 40 factors (accounting for less than 2% of the total number of factors/terms in each case) of the intermediary literature sets *Adenosine Triphosphate*, *Norepinephrine*, *Protein Kinases*, *Propranolol*, and *Verapamil*. In each example of the intermediaries, evd brings source and target descriptor in (sometimes very) close vicinity (Table 8).

Table 7 Factor analysis of the *Cardiac Hypertrophy* source literature set.
Retrieval details: PubMed title search for **"heart enlargement" OR "cardiac hypertrophy" OR "heart hypertrophy" OR "enlarged heart" OR "hypertrophic cardiomyopathy" OR "left ventricular hypertrophy" OR "right ventricular hypertrophy"**, limited to all publication years up to and including 2003.
Retrieval date: October 19, 2005.
Size of the document-by-term matrix: $7,510 \times 2,247$.
Source terms: *Cardiomegaly* (CM); *Cardiomyopathy, Hypertrophic* (CH); *Cardiomyopathy, Dilated* (CD); *Hypertrophy, Left Ventricular* (HLV); *Hypertrophy, Right Ventricular* (HRV).
Intermediary terms: *Norepinephrine* (Nor), *Verapamil* (Ver), *Propranolol* (Pro), *Protein Kinases* (PK), *Adenosine Triphosphate* (ATP).
Other details: see Table 1

Factor	Position of source term	Position of intermediary term	Distance
1	88 (CD)	104 (Nor)	15
		93 (Pro)	4
2	42 (HLV)	56 (Nor)	13
	99 (HRV)	100 (ATP)	0
3	74 (CD)	88 (Pro)	13
4	783 (CD)	766 (Ver)	16
5	113 (CD)	124 (Ver)	10
8	38 (Cm)	44 (Ver)	5
	273 (CD)	289 (PK)	15
	123 (HRV)	109 (Nor)	13
9	63 (CD)	78 (Ver)	14
10	86 (HRV)	85 (Nor)	0
11	285 (HRV)	281 (Nor)	3
14	98 (CD)	78 (Pro)	19
15	30 (Cm)	44 (Ver)	13
		49 (Pro)	18
	67 (CH)	49 (Pro)	17
16	128 (CD)	111 (Nor)	16
		115 (Pro)	12
		126 (ATP)	1
19	116 (HRV)	127 (Pro)	10
20	209 (HRV)	207 (ATP)	1

Table 8 Factor analysis of literature sets intermediary to *Cardiac Hypertrophy* and *Chlorpromazine*.
Retrieval details, Retrieval dates; Size of document-by-term matrices:
PubMed title searches, limited to all publication years up to and including 2003.
Adenosine Triphosphate (ATP): **adenosine triphosphate**, October 18, 2005; $2,752 \times 2,296$.
Norepinephrine (Nor): **norepinephrine**, February 14, 2005; $3,770 \times 2,859$.
Protein Kinases (PK): **protein kinases**, October 18, 2005; $3,770 \times 2,859$.
Propranolol (Pro): **propranolol**, October 18, 2005; $7,679 \times 2,824$.
Verapamil (Ver): **verapamil**, October 18, 2005; $5,325 \times 2,386$.
Descriptor extraction, eigenvalue decompositon, factor screening: see Sect. 2.
Short source–target distance: 40 or less terms between one of the *Cardiac Hypertrophy* source terms (see Table 7) and the target term *Chlorpromazine* (*in parentheses*: number of distances of 10 or less terms).
The first 40 factors derived from each intermediary literature set were screened

Intermediary literature set	Number of short source–target distances
ATP	3 (1)
Nor	6 (3)
PK	2 (–)
Pro	11 (5)
Ver	7 (2)

3.4 Multiple Sclerosis

Erythropoietin might act as neuroprotective agent and thus be beneficial to *Multiple Sclerosis* [31, 32]. In September 2005 we retrieved 14 records mentioning *Multiple Sclerosis* and *Erythropoietin* from the whole PubMed database. The four papers published prior to 2001 do not focus at a direct relationship between the disease and the substance (as infered from the papers' abstracts), and to none of these four records the MeSH descriptor *Multiple Sclerosis* has been assigned. Therefore, the pair *Multiple Sclerosis* – *Erythropoietin* may serve as yet another good example of CBD literatures, and we applied the decomposition method to the source *Multiple Sclerosis* literature. We searched PubMed for documents with the title phrase "multiple sclerosis", and downloaded the nearly 4,100 records with publication dates from 1995 to 2000, extracted the MeSH descriptors and subjected the document-by-term matrix to evd as described in Sect. 2. Within the first 30 factors we found two instances of very short distances (ten terms or less) between the source descriptor *Multiple Sclerosis* and a potential intermediary term, *Nitric Oxide Synthase*, (Table 9). The respective analysis of the intermediary *Nitric Oxide Synthase* literature resulted in one factor (within the first 30) with close vicinity of the source term *Multiple Sclerosis* and the target descriptor *Erythropoietin* (Table 10). The relative descriptor frequency is obviously at the lower limit: 0.1 and 0.2% of the *Nitric Oxide Synthase* literature for *Erythropoietin* and *Multiple Sclerosis*, respectively. The descriptor *Nitric Oxide Synthase* occurs in 0.4% of the *Multiple Sclerosis* literature set.

Table 9 Factor analysis of the *Multiple Sclerosis* source literature set.
Retrieval details: PubMed title search for **multiple sclerosis**, limited to publication years 1995–2000.
Retrieval date: September 22, 2005.
Size of the document-by-term matrix: $4,041 \times 2,012$.
Source term: *Multiple Sclerosis.*
Intermediary term: *Nitric-Oxide Synthase.*
Other details: see Table 1

Factor	Position of source term	Position of intermediary term	Distance
24	91	102	10
28	229	219	9

Table 10 Factor analysis of the *Nitric-Oxide Synthase* intermediary literature set.
Retrieval details: PubMed title search for **"nitric oxide synthase"**, limited to publication years 1995–2000.
Retrieval date: September 27, 2005.
Size of the document-by-term matrix: $4,889 \times 3,169$.
Source term: *Multiple Sclerosis.*
Target term: *Erythropoietin.*
Other details: see Table 1

Factor	Position of source term	Position of target term	Distance
11	584	573	10

This final example of complementary-but-disjoint literatures makes clear that the detection of possible path(s) from source via intermediary(ies) to target(s) may be rather difficult due to low frequencies of relevant terms and short distances but also demonstrates the sensitivity of the factor-analytic method, provided a reasonable hypothesis is hidden in the literature data.

4 Discussion

Reduction of the semantic space by matrix decomposition as a tool in literature-based detection of indirect links with the potential of new hypotheses has first been applied by Gordon and Dumais using words and phrases from titles and abstracts as representations of the appropriate literature sets [13]. The MeSH descriptors assigned to PubMed records by PubMed indexers are taken from a comprehensive controlled vocabulary and also represent the content of the underlying articles. In the SL experiments reported here we factorized the semantic space constituted by MeSH terms and inspected the resulting eigenvectors directly. We were able to show that – while restricting the number of factors to be screened to a small fraction starting with the first factor – terms relevant to the discovery process (i.e. source and intermediary and source and target terms) can be found close to each other in several factors. The method works with small and large literature sets, ranging from several hundred to several thousand documents and MeSH descriptors. It is important that the method acts reliably on both instances: to detect intermediary concepts in source literature, and target concepts in intermediary literature. Moreover, screening factors derived from target literature sets finds known target and intermediary descriptors in close vicinity within the limits (relating to number of factors and terms to be screened) used throughout our investigations (data not shown).

Of crucial importance in literature-based discovery is, of course, the "human factor": experts have to study term lists of whatever origin and to select promising terms. Screening of long lists is certainly tedious but we think that screening within our chosen limits of 30–40 factors/terms is tolerable and does not exhaust the human inspector. On the basis of the results presented we conclude that relevant intermediary and target terms in relation to known source terms can be detected under those constraints in most cases.

The matrix decomposition feature could be implemented into existing text mining tools as *Charité MLink* [26, 27], and offer an additional and alternative way of term-screening to the user.

5 Conclusion

A method was described which uses factorised MeSH term matrices in order to find by quick and easy manual screening terms close to each other which might be of interest in the process of linking disparate but complementary literatures. The method supplements the already existing approaches to literature-based hypothesis generation.

Acknowledgements The main parts of our work described here have been performed in 2005 when J.S. was still an employee in the Medical Library at the Campus Benjamin Franklin of the Charité, Berlin, Germany. Preliminary results have been presented in part at the 10th International Conference of the International Society for Scientometrics and Informetrics, Stockholm, July 24–28, 2005 [36].

We thank Prof. Dr. Thomas Tolxdorff and Dr. Juergen Braun, head and vice-head, respectively, of the Institute of Medical Informatics of the Charité for continuous support of our work.

This work was supported by the Deutsche Forschungsgemeinschaft, grant no. LIS 4-542 81.

References

1. Swanson, D.R.: Fish oil, Raynaud's syndrome, and undiscovered public knowledge. Perspectives in Biology and Medicine **30** (1986) 7–18
2. Swanson, D.R.: Migraine and magnesium: eleven neglected connections. Perspectives in Biology and Medicine **31** (1988) 526–557

3. Swanson, D.R.: Online search for logically-related noninteractive medical literatures: a systematic trial-and-error strategy. Journal of the American Society for Information Science **40** (1989) 356–358
4. Swanson, D.R.: A second example of mutually isolated medical literatures related by implicit, unnoticed connections. Journal of the American Society for Information Science **40** (1989) 432–435
5. Swanson, D.R.: Somatomedin c and arginine: implicit connections between mutually isolated literatures. Perspectives in Biology and Medicine **33** (1990) 157–186
6. Stegmann, J., Grohmann, G.: Hypothesis generation guided by co-word clustering. Scientometrics **56** (2003) 111–135
7. Swanson, D.R., Smalheiser, N.R.: Implicit text linkages between medline records: using Arrowsmith as an aid to scientific discovery. Library Trends **48** (1999) 48–59
8. Kostoff, R.N.: Science and technology innovation. Technovation **19** (1999) 593–604
9. Smalheiser, N.R., Swanson, D.R.: Indomethacin and Alzheimer's disease. Neurology **46** (1996) 583
10. Smalheiser, N.R., Swanson, D.R.: Linking estrogen to Alzheimer's disease: an informatics approach. Neurology **47** (1996) 809–810
11. Smalheiser, N.R., Swanson, D.R.: Calcium-independent phospholipase A_2 and schizophrenia. Archives of General Psychiatry **55** (1998) 752–753
12. Gordon, M.D., Lindsay, R.K.: Toward discovery support systems: a replication, re-examination and extension of Swanson's work on literature-based discovery of a connection between Raynaud's and fish oil. Journal of the American Society for Information Science **47** (1996) 116–128
13. Gordon, M.D., Dumais, S.: Using latent semantic indexing for literature based discovery. Journal of the American Society for Information Science **49** (1998) 674–685
14. Weeber, M., Klein, H., de Jong-van den Berg, L.T.W., Vos, R.: Using concepts in literature-based discovery: simulating Swanson's raynaud-fish oil and migraine-magnesium discoveries. Journal of the American Society for Information Science and Technology **52** (2001) 548–557
15. Srinivasan, P.: Text mining: generating hypotheses from medline. Journal of the American Society for Information Science and Technology **55** (2004) 396–413
16. Stegmann, J., Grohmann, G.: Advanced information retrieval for hypothesis generation. Journal of Information Management and Scientometrics **1** (2004) 46–53
17. Stegmann, J., Grohmann, G.: Transitive text mining for information extraction and hypothesis generation. arXiv:cs.IR/0509020 (2005)
18. The Unified Medical Language System: http://www.nlm.nih.gov/mesh/presentations/tafumls/index.htm
19. Medical Subject Headings: http://www.nlm.nih.gov/mesh/
20. Bekhuis, T.: Conceptual biology, hypothesis discovery, and text mining: Swanson's legacy. Biomedical Digital Libraries **3** (2006) Article No. 2
21. Wren, J.D., Bekeredjian, R., Stewart, J.A., Shohet, R.V., Garner, H.R.: Knowledge discovery by automated identification and ranking of implicit relationships. Bioinformatics **20** (2004) 389–398
22. Hristovski, D., Stare, J., Peterlin, B., Dzeroski, S.: Supporting discovery in medicine by association rule mining in Medline and UMLS. Medinfo **2** (2001) 1344–1348
23. Perez-Iratxeta, C., Bork, P., Andrade, M.A.: Association of genes to genetically inherited diseases using data mining. Nature Genetics **31** (2002) 316–319
24. Hristovski, D., Peterlin, B., Mitchell, J.A., Humphrey, S.M.: Improving literature based discovery support by genetic knowledge integration. Studies in Health Technology and Informatics **95** (2003) 68–73
25. Callon, M., Courtial, J.P., Laville, F.: Co-word analysis as a tool for describing the network of interactions between basic and technological research: the case of polymer chemistry. Scientometrics **22** (1991) 155–205
26. Charité MLink: http://mlink.charite.de/

27. Grohmann, G., Stegmann, J.: C-mlink: a web-based tool for transitive text mining. In Ingwersen, P., Larsen, B., eds.: Proceedings of ISSI 2005 – 10th International Conference of the International Society for Scientometrics and Informetrics: 24–28 July 2005; Stockholm, Sweden, Karolinska University Press, Stockholm (2005), pp. 658–659
28. Deerwester, S., Dumais, S.T., Furnas, G.W., Landauer, T.K., Harshman, R.: Indexing by latent semantic analysis. Journal of the American Society for Information Science **41** (1990) 391–407
29. Homayouni, R., Heinrich, K., Wei, L., Berry, M.W.: Gene clustering by latent semanting indexing of Medline abstracts. Bioinformatics **21** (2005) 104–115
30. Supplemental online material for the paper [21]: http://innovation.swmed.edu/IRIDESCENT/Supplemental_Info.htm
31. Ehrenreich, H., Aust, C., Krampe, H., Jahn, H., Jacob, S., Herrmann, M., Siren, A.L.: Erythropoietin: novel approaches to neuroprotection in human brain disease. Metabolic Brain Disease **19** (2004) 195–206
32. Diem, R., Sattler, M.B., Merkler, D., Demmer, I., Maier, K., Stadelmann, C., Ehrenreich, H., Bahr, M.: Combined therapy with methylprednisolone and erythropoietin in a model of multiple sclerosis. Brain **128** (2005) 375–385
33. PubMed: http://www.ncbi.nlm.nih.gov/entrez/query.fcgi?db=PubMed
34. R-DEVELOPMENT-CORE-TEAM: R: a language and environment for statistical computing. R Foundation for Statistical Computing, Vienna (2004)
35. Kostoff, R.N., Block, J.A., Stump, J.A., Pfeil, K.M.: Information content in Medline record fields. International Journal of Medical Informatics **73** (2004) 515–527
36. Stegmann, J., Grohmann, G.: Factor analytic approach to transitive text mining using Pubmed keywords. In Ingwersen, P., Larsen, B., eds.: Proceedings of ISSI 2005 – 10th International Conference of the International Society for Scientometrics and Informetrics: 24–28 July 2005; Stockholm, Sweden, Karolinska University Press, Stockholm (2005), pp. 280–283

Literature-Based Knowledge Discovery using Natural Language Processing

D. Hristovski, C. Friedman, T.C. Rindflesch, and B. Peterlin

Abstract Literature-based discovery (LBD) is an emerging methodology for uncovering nonovert relationships in the online research literature. Making such relationships explicit supports hypothesis generation and discovery. Currently LBD systems depend exclusively on co-occurrence of words or concepts in target documents, regardless of whether relations actually exist between the words or concepts. We describe a method to enhance LBD through capture of semantic relations from the literature via use of natural language processing (NLP). This paper reports on an application of LBD that combines two NLP systems: BioMedLEE and SemRep, which are coupled with an LBD system called BITOLA. The two NLP systems complement each other to increase the types of information utilized by BITOLA. We also discuss issues associated with combining heterogeneous systems. Initial experiments suggest this approach can uncover new associations that were not possible using previous methods.

1 Introduction

Literature-based discovery (LBD) is a method for automatically generating hypotheses for scientific research by finding overlooked implicit connections in the research

D. Hristovski
Institute of Biomedical Informatics, Medical Faculty, University of Ljubljana, Vrazov trg 2/2, 1104 Ljubljana, Slovenia
dimitar.hristovski@mf.uni-lj.si

C. Friedman
Department of Biomedical Informatics, Columbia University, 622 West 168 St, New York, NY 10032, USA

T.C. Rindflesch
National Library of Medicine, Bethesda, Maryland, USA

B. Peterlin
Division of medical genetics, UMC, Slajmerjeva 3, Ljubljana, Slovenia

P. Bruza and M. Weeber (eds.), *Literature-based Discovery,*
Springer Series in Information Science and Knowledge Management 15.

literature. Discoveries have the form of relations between two primary concepts, for example a drug as a treatment for a disease or a gene as the cause of a disease. Swanson [1] introduced a paradigm in which such relations are discovered in bibliographic databases by uncovering a third concept (such as a physiologic function) that is related to both the drug and the disease. The discovery of the third concept allows a relation between the primary concepts, which was latent in the literature, to become explicit, thus constituting a potential discovery.

Current literature-based discovery systems (for example [2–12] use concept co-occurrence as their primary mechanism. No semantic information about the nature of the relation between concepts is provided. The use of co-occurrence has several drawbacks, since not all co-occurrences underlie "interesting" relations: (a) users must read large numbers of Medline citations when reviewing candidate relations; (b) systems tend to produce large numbers of spurious relations; and, finally, (c) there is no explicit explanation of the discovered relation.

In this chapter we address these deficiencies by enhancing the literature-based paradigm with the use of semantic relations to augment co-occurrence processing. We combine the output of two natural language processing systems to provide these predications: SemRep [13] and BioMedLee [14]. On the basis of explicit semantic predications, the user can ignore relations which are either uninteresting (thus reducing the amount of reading required) or wrong (eliminating false positives). Analysis using predications can support an explanation of potential discoveries.

2 Background

2.1 Literature-Based Discovery

The methodology in literature-based discovery relies on the notion of concepts relevant to three literature domains: X, Y, and Z. In a typical scenario, X concepts are those associated with some disease and Z concepts relate to a drug that treats the disease. Y concepts might then be physiological or pathological functions, symptoms, or body measurements. Concepts in X and Y are often discussed together, as are those in Y and Z. However, concepts from X and Z may not appear together in the same research paper. Discovery is facilitated by using particular Y concepts to draw attention to a connection between X and Z that had not been previously noticed.

In implementation, all the Y concepts in a bibliographic database related to the starting concept X are usually computed first. Then the Z concepts related to Y are found. Those Y concepts that appear with both X and Z provide the link from X to Z. The user then checks whether X and Z appear together in the research literature; if they do not, a potentially useful relation has been discovered. This relation needs to be confirmed or rejected using human judgment, laboratory methods, or clinical investigations.

In a discovery reported by Swanson [1], the X domain was Raynaud's disease. Of the many Y terms co-occurring with this disorder, blood viscosity and platelet aggregation were found to co-occur with a Z term, fish oil (rich in eicosapentaenoic acid). Fish oil (Z) reduces blood viscosity and platelet aggregation (Y), which are increased in Raynaud's disease (X), and thus fish oil was proposed as a new treatment for Raynaud's disease. Swanson has published several other medical discoveries using this methodology. However, in his original work (and in all subsequent replications of this discovery), what is increased in relation to the disease and what can be used to decrease it, must be determined by reading relevant Medline citations. This is exactly where we want to improve the state-of-the-art in LBD.

Several methods are being pursued in current LBD systems (for a more detailed review see [15]). Some systems extract concepts from the titles and abstracts of Medline citations (often using MetaMap [16]), while others use the assigned MeSH descriptors to represent concepts in citations. All systems use co-occurrence to determine which concepts are in a relationship, although some augment co-occurrence with other derived relation measures. Usually the semantic types of the concepts are used to filter out unneeded relations and concepts.

Swanson and Smalheiser have developed a system called Arrowsmith [2], which uses co-occurrence of words or phrases from the title of Medline citations. The BITOLA system (Hristovski et al. [3,4]) uses association rules as a relation measure between concepts. In general, association rule mining [17] finds interesting associations and/or correlation relationships among large set of data items. In BITOLA a data item corresponds to a Medline citation and is represented as a set of concepts. For each citation, the concepts are the assigned MeSH headings and additionally gene symbols extracted from the titles and abstracts of Medline citations. For example, the association rule *Multiple Sclerosis* → *Optic Neuritis* tells us that there is probably some association between *Multiple Sclerosis* and *Optic Neuritis*, but does not tell us the semantic nature of this association.

Weeber et al. [5] use MetaMap to identify UMLS concepts in titles and abstracts and use concept co-occurrence as a relation measure. For filtering, they use UMLS semantic types. For example, the semantic type of one of the co-occurring concepts might be set to *Disease or Syndrome* and the other to *Pharmacologic Substance*, thus only co-occurrences between a disease and a drug are found. Lindsay and Gordon [6] use an approach similar to Arrowsmith but add various information retrieval techniques to assign weight to the terms being manipulated. Gordon and Dumais [7] employ a statistical method called latent semantic indexing to assist in LBD. Wren [8] uses mutual information measures for ranking target terms based on their shared associations. Srinivasan [12] developed a system, called Manjal, which uses MeSH terms as concepts and term weights instead of simple term frequencies. For ranking, the system uses an information retrieval measure based on term co-occurrence. Pratt [9] uses MetaMap to extract UMLS concepts from the titles of Medline citations and then uses association rules as a relationship measure between concepts.

The Telemakus system [10] is different from the rest of the systems mentioned in so far as it uses manually extracted relationships to represent the research findings. Each relationship is a pair of concepts from the article's figure and title legends.

The semantic relation between the concepts is not extracted. The manual relation extraction method has two consequences: the positive one is that the method has high precision and the negative one is that it is time consuming and thus currently used in only two relatively narrow domains.

Recently Hu [11] presented a system called Bio-SbKDS where MeSH terms are used as concepts. This system uses the relations between semantic types from the UMLS Semantic Network for two purposes: to filter out uninteresting concepts, and to guess the semantic relation between concepts. In other words, if two concepts co-occur in a Medline citation, the relation between the corresponding semantic types of these two concepts is used as the semantic relation between the concepts. This is only an approximation because there is no guarantee that if the concepts co-occur they are semantically related and also there is ambiguity in the UMLS Semantic Network because often more then one semantic relation is present between two semantic types. However, this approach seems to work quite well in replicating Swanson's Raynaud's – fish oil discovery.

Our method differs from all the above methods because we use natural language processing (NLP) techniques to augment co-occurrences with specific types of relations, which are obtained as a result of using two different NLP systems.

2.2 Natural Language Processing

2.2.1 BioMedLEE Natural Language Processing System

BioMedLEE captures genotypic and phenotypic information and relations from the literature, and is a recent adaptation of MedLEE [18, 19], which was developed to structure and encode telegraphic clinical information in the patient record. BioMedLEE is based on a symbolic grammar formalism that combines syntax and semantics, using a lexicon to specify semantic and syntactic classes for words and phrases in the domain. The lexicon consists of a modified and augmented version of MedLEE's lexicon, which was derived from clinical documents, the UMLS (Unified Medical Language System) [20], and other online biomedical knowledge sources, but this work focuses on use of the concepts that correspond to UMLS Metathesaurus concepts only. BioMedLEE consists of a number of different text processing modules, each of which aims to regularize specific aspects of text processing while minimizing loss of information. The following is a brief summary of the primary modules and the resources they use:

a. *Abbreviation and Parenthesis Component*: This module identifies abbreviations explicitly defined in the article, and tags them so that the subsequent modules will be able to substitute the full form in place of the abbreviation. For example, HD, in *Huntington Disease (HD)* will be assumed to be *Huntington Disease* throughout the article. Other parenthetical expressions may be tagged so that they will be ignored during parsing.

b. Biomolecular Named Entity Recognition and Normalization: This module uses part of speech tagging to recognize the boundaries of noun phrases, and then identifies ones that appear to be biomolecular entities, such as the names of genes, gene products, and other substances. The terms that are biomolecular entities are then matched against a database of biomolecular entities using regular expressions that allow for certain variations (e.g. *il-2, il 2, il2*). When a match is found, the term is tagged so that the tag includes the semantic category (e.g. gene/gene product, substance), and the target output form. For example, after tagging is performed (we assume here that the tagging module used a database of UMLS genes and proteins to normalize biomolecular entities), the tagged output for the sentence "*Axonal transport of N-terminal huntingtin suggests pathology of corticostriatal projections associated with HD*" will be "*Axonal transport of N-terminal <phr sem="gp"t="UMLS: C1415504_hd gene"> huntingtin</phr> suggests pathology of corticostriatal projections associated with <phr sem="disease" t="Huntington's disease">HD</phr>*". The tag around *huntingtin* has an attribute, which is a semantic category **sem** with value **gp** representing the category **gene/gene product** and a target form attribute **t**, which, in this case, is the UMLS code previously generated by the tagger. In addition, there is a tag around, *HD*, with a semantic category **disease** and target form **Huntington's disease**, which is the full form that occurred previously in the article along with the abbreviation *HD*.

c. Preprocessing Component: This module determines section and sentence boundaries, and performs lexical lookup for the remaining parts of the sentence that were not tagged in b. above. This would include phenotypic entities, such as anatomical locations, diseases, and processes, as well as functional English words. For example, "corticostriatal" would be identified as an anatomical concept, and "suggest" would be identified as a relation that could connect two biomedical entities. The relations are semantic relations that have been categorized based on linguistic characteristics and are not necessarily UMLS relations.

d. Parser: This module extracts, structures, and encodes phenotypic and genotypic entities and relations for tagged text from the previous module using a grammar and a lexicon to parse and structure the output, and a coding table to map the normalized output to ontological codes. The output is in an XML form based on a representational schema of the domain, called PGschema [21], which represents genotypic and phenotypic entities, their ontological codes, modifiers, and relations between the entities. Figure 1 shows an example of a simplified output form generated by BioMedLEE for the above tagged sentence, where some of the nested tags have been manually indented to facilitate readability of the output structure. This output differs from output generated by systems that use co-occurrence of terms because BioMedLEE found actual relations "suggest" and "associated with" in the text. The relation "suggest" connects "axonal transport of hd gene" with a second nested relation "associated with", whose first argument is "pathology" with an anatomical modifier "corticostriatal" and whose second argument is "Huntington's disease".

```
<relation v = "suggest">
  <bodyfunc v = "transport"><arg v = "1"></arg>
    <bodyloc v="axon"></bodyloc><cellcomp v ="N-terminal"></cellcomp>
    <gene_gproduct v = "UMLS: C1415504_hd gene" idref="p126">
    </gene_gproduct>
    <code v ="UMLS:C0004462_axonal transport"</code>
  </bodyfunc>
  <relation v = "associated with"><arg v = "2"></arg>
    <problem v = "pathology"><arg v = "1"></arg>
    <bodyloc v = "corticostriatal"></bodyloc>
    </problem>
    <problem v = "Huntington's disease"><arg v = "2"></arg>
      <code v = "UMLS:C0020179_huntington disease"></code>
    </problem>
  </relation>
</relation>
```


Fig. 1 Simplified XML output generated by BioMedLee for a sample sentence

2.2.2 SemRep Natural Language Processing System

SemRep [13] is a symbolic natural language processing system for identifying semantic predications in biomedical text. The current focus is on Medline citations. Linguistic processing is based on an underspecified (shallow) parse structure supported by the SPECIALIST Lexicon [22] and the MedPost part-of-speech tagger [23]. Medical domain knowledge is provided by the UMLS. Predications produced by SemRep consist of Metathesaurus concepts as arguments of a Semantic Network relation.

For this project, the most important relation is TREATS; however, SemRep identifies additional semantic predications representing various aspects of biomedicine. The core relations addressed refer to clinical actions (e.g. TREATS, PREVENTS, ADMINISTERED_TO, MANIFESTATION_OF) and organism characteristics (LOCATION_OF, PART_OF, PROCESS_OF). SemRep has recently been enhanced to address pharmacogenomics text [24]. Relations in this semantic area refer to substance interactions and pharmacologic effects (AFFECTS, CO-EXISTS_WITH, DISRUPTS, AUGMENTS, INTERACTS_WITH, INHIBITS, STIMULATES), as well as genetic etiology (ASSOCIATED_WITH, PREDISPOSES, CAUSES). The majority of SemRep's relations are drawn from the Semantic Network; however, several have been defined to extend the coverage of that ontology, including ADMINISTERED_TO, CO-EXISTS_WITH, and PREDISPOSES.

Each semantic relation serves as the predicate of an ontological predication that controls SemRep processing. The arguments in these predications are UMLS semantic types, such as 'Human' or 'Anatomical Structure', which can, for example, appear in the predication "Anatomical Structure PART_OF Human." All predications extracted from text by SemRep must conform to an ontological predication.

Semantic interpretation is based on the underspecified parse structure, in which simple noun phrases are enhanced with corresponding Metathesaurus concepts by

1) [[head(noun(treatment)), metaconc('Treatment':[topp]))], [prep(of)], [head(noun([huntington's disease)), metaconc('Huntington Disease':[dsyn]))], [prep(with)], [head(noun([amantadine)), metaconc('Amantadine':[orch,phsu]))]]

2) 'Pharmacological Substance' TREATS 'Disease or Syndrome'

3) Amantadine TREATS Huntington Disease

Fig. 2 SemRep processing of treatment of Huntington's disease with amantadine

MetaMap [16]. For example, processing of the phrase *treatment of Huntington's disease with amantadine* produces the structure seen in (1) in Fig. 2. The noun phrase *Huntington's disease* has been mapped to the concept "Huntington's disease," with semantic type 'Disease or Syndrome' (dsyn).

The parse structure enhanced with Metathesaurus concepts serves as the basis for the final phase in constructing a semantic predication. During this phase, SemRep applies "indicator" rules which map syntactic elements (such as verbs and nominalizations) to the predicate of an ontological predication. Argument identification rules (which take into account coordination, relativization, and negation) then find syntactically allowable noun phrases to serve as arguments for indicators. If an indicator and the noun phrases serving as its syntactic arguments can be interpreted as a semantic predication, the following condition must be met: The semantic types of the Metathesaurus concepts for the noun phrases must match the semantic types serving as arguments of the indicated ontological semantic predication. For example, in Fig. 2 *treatment* is an indicator for TREATS, with the corresponding ontological predication seen in (2) in Fig. 2. The concepts corresponding to the noun phrases *amantadine* and *Huntington's disease* can serve as arguments of TREATS because their semantic types ('Pharmacological Substance' (phsu) and 'Disease or Syndrome' (dsyn)) match those in the ontological predication. In the final interpretation, (3) in Fig. 2, the Metathesaurus concepts from the noun phrases are substituted for the semantic types in the ontological predication.

3 Methods

3.1 Discovery Patterns

3.1.1 The Relations Maybe_Treats1 and Maybe_Treats2

In order to exploit semantic predications in literature-based discovery, we introduce the notion of a *discovery pattern*, which contains a set of conditions to be satisfied for the discovery of new relations between concepts. The conditions are combinations of relations between concepts extracted from Medline citations. In this paper we deal with the *Maybe_Treats* pattern, which has two forms: *Maybe_Treats1* and *Maybe_Treats2* (Fig. 4). In both forms the goal is to propose potential new treatments, and the two can work in concert: proposing either two different new treatments (complementarity) or the same treatment by using different discovery reasoning (reinforcement). The following reasoning is used as a novelty check for the proposed new treatments (stated informally in terms of the X, Y, Z paradigm): It is a discovery that drug Z maybe treats disease X if there is currently no evidence in the medical literature that drug Z is already used to treat disease X.

The two discovery patterns are different in the way they generate new candidate treatments Z. The first form *Maybe_Treats1* is satisfied when there is a change in a substance, body function, or body measurement (concept Y) associated with the starting disease X, and there is also an opposite change in concept Y associated with concept Z. In other words, we first try to find the characteristics of a disease X with regard to a change in the level of substance or measurement Y in patients with this disease. Then we look for a drug or chemical Z that can cause an opposite change in the same substance or measurement Y. That is, if the Y concept decreases in association with the X disease, we expect it to increase in association with the Z drug, or vice versa. An example of the first form is the reasoning used by Swanson to propose fish oil (Z) as a new treatment for Raynaud's disease (X). Fish oil (Z) was proposed because it reduces blood viscosity (Y) which was reported in the literature to be increased in patients with Raynaud's.

In using *Maybe_Treats2* to find a potential new treatment for a starting disease X we first search for another disease X2 that has characteristics similar to X (Y2 substance or function is either increased or decreased in both X and X2). Then we propose as a new treatment for disease X the drug (Z2) already used to treat disease X2, if there is no evidence in the literature that Z2 is already used to treat X. An example of this might be what we have observed while performing this research. In patients with Huntington disease (HD) the level of insulin is often decreased. The level of insulin is also decreased in diabetes mellitus (type 1). Therefore, treatments for diabetes might also be used for HD.

We can formally define the two forms of the *Maybe_Treats* discovery pattern using the predications in Figs. 3 and 4.

3.1.2 The Relations Associated_with_change and Treats

The relations *Associated_with_change* and *Treats* are used to extract known facts from the biomedical literature. The relations *Maybe_Treats1* and *Maybe_Treats2* predict potentially new treatments based on the known facts extracted by *Associated_with_change* and *Treats*. *Associated_with_change* is used to extract a relation in which one concept is associated with a change in another concept (e.g. a disease associated with an increase in the level of a substance). For the extraction of *Associated_with_change* we use BioMedLee. The relation *Treats* is used to extract drugs known to treat a disease according to the literature. The major purpose of this relation in our approach is to eliminate the drugs already known to be used for treatment

```
Maybe_Treats(Drug_Z, Disease_X)  IF
        Maybe_Treats1(Drug_Z, Disease_X)  OR
        Maybe_Treats2(Drug_Z, Disease_X).

Maybe_Treats1(Drug_Z, Disease_X) IF
        Associated_with_change(Disease_X,Subst_Y,Change_Y11) AND
        Associated_with_change (Drug_Z, Subst_Y,Change_Y12)   AND
        Opposite_Change(Change_Y11, Change_Y12)    AND
        NOT Treats(Drug_Z, Disease_X).
Opposite_Change("Increase", "Decrease").
Opposite_Change("Decrease", "Increase").

Maybe_Treats2(Drug_Z2, Disease_X) IF
        Associated_with_change (Disease_X,Subst_Y2,Change_Y21)  AND
        Associated_with_change (Disease_X2,Subst_Y2,Change_Y22) AND
        Same_Change(Change_Y21,Change_Y22)  AND
        Treats(Drug_Z2, Disease_X2) AND
        NOT Treats(Drug_Z2, Disease_X).
Same_Change("Increase", "Increase").
Same_Change("Decrease", "Decrease").
```

Fig. 3 Formal definition of the discovery pattern *Maybe_Treats*

from the list of drugs or chemicals that have not been used, but seem promising. Additionally, in the *Maybe_Treats2* form, the *Treats* relation is used to find existing treatments to similar diseases. *Treats* relations are identified by SemRep.

The relation *Associated_with_change* is a higher level relation and is based on basic BioMedLee relations. In this research, we used three methods to derive *Associated_with_change* where the first two are the most credible. The first is based on the binary *Increase* or *Decrease* relations. For example, for the sentence "*Speech production increases cerebral blood flow* in HD patients", BioMedLEE extracts *Increase(Speech production, cerebral blood flow)*. In this example, although the binary relation associated with "increase" was extracted, the relation "in HD patients" was lost because BioMedLEE did not recognize that the abbreviation HD referred to Huntington's disease.

The second method is to use binary relations in which one of the arguments has a change such as *Increase* or *Decrease* associated directly with the argument. The relation can be any of those that indicate some kind of an association between its arguments, such as *associated_with, exhibited, due to, suggest, results from.* For example, from the sentence "*Huntington's disease* brains all *exhibited* a marked *decrease* in *substance P* fiber density in the substantia nigra and globus pallidus" BioMedLee extracts *Exibit(Huntington disease, Substance P/decrease)*.

The third way to derive *Associated_with_change* relations is to exploit phrase or sentence level co-occurrence of concepts with which a change is associated with one of the concepts. In other words, we extract all the concepts from a phrase or sentence and if there is at least one concept with a change directly associated with it, we then assume that that concept is related to the other concepts in the same phrase or sentence. Obviously, this is the least credible way of deriving *Associated_with_change* relations; however, it significantly improves recall. For example,

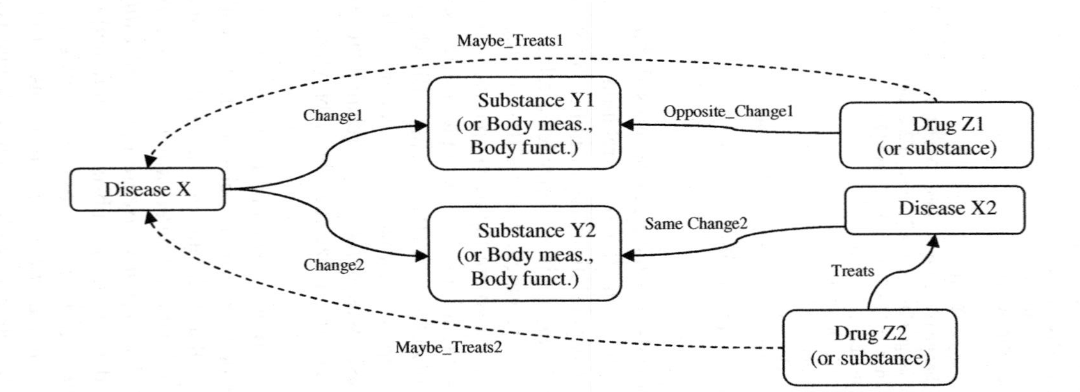

Fig. 4 Discovery pattern *Maybe_Treats*. *Maybe_Treats1* proposes Z1 (drug or substance) as a new treatment for disease X because Z1 causes opposite change to Y1 (function or substance) and the change of Y1 is a characteristic of disease X. *Maybe_Treats2* proposes Z2 as a new treatment of X because there is a similar disease, X2, and drug Z2 is known to treat X2

from the sentence "In *Huntington's disease*, there is a *decrease* of the *neuropeptides*, substance P, enkephalins, and cholecystokinin in the striatonigral system, whereas in *Parkinson's disease* an *increase* of substance P is found in the substantia nigra", BioMedLee extracts *co_occurs(Huntington's disease, Neuropeptides/decrease)* which is correct, but from the same sentence the system also extracts *co_occurs (Parkinson's disease/increase, Huntington's disease)*, which is not correct.

It is possible to use the *Maybe_Treats* pattern (both forms) for several discovery tasks depending on what input is provided. If a drug Z is provided as input, the pattern will try to generate diseases X that might be treated. If a disease X is provided as input, the pattern will try to generate drug Z that might be used to treat the disease X. If both a disease X and a drug Z are provided as input, the pattern will test whether the drug might be used to treat the disease. If it can, the pattern can generate an explanation through the intermediate concepts Y. For example, the drug Z might be used to treat X because Y is increased in disease X, and Z has been reported to decrease the level of Y.

3.2 Integrated BioMedLEE and SemRep Output Format

The output formats normally provided by BioMedLee and SemRep are different from each other, and therefore it was not straightforward to combine the use of both systems. To enable the integration of the output of the two systems for the purpose of this research, we developed a *common output format*, the specification of which is still evolving. Currently, the common format contains three types of lines: *text*, *entity* and *relation*. Each type of line is a delimited list of fields. The input to both systems is a set of Medline citations. Each Medline citation is broken into a sequence of sentences and each sentence is processed separately. For each sentence, a line of type *text* is first generated to present the actual text of the current sentence. Then a line of type *entity* is generated for each biomedical entity (concept) extracted from the current sentence regardless of whether the entity is part of a relation or not. Finally, all the relations between the entities from the current sentence are generated as lines of type *relation*.

Table 1 shows the fields used in the common format. All three types of lines start with fields 1–6. The first field is the system identification to indicate which system generated the line, the second is the PubMed identification number, followed by the subsection abbreviation, which indicates whether the sentence comes from the methods, conclusions, results or some other subsection of a structured abstract. The fourth field specifies whether the sentence is from the title or the abstract. The fifth field specifies the sentence identification, which is slightly different for each system because different methods are used to recognize sentence boundaries. The sixth field identifies the row type, which is one of *text*, *entity* or *relation*. This field determines the format of the rest of the line.

For a line of type *text*, the next field is the actual text of the sentence, which for BioMedLee is in a tagged text format where the tags are linked to the entities,

Table 1 The common format used to represent BioMedLee and SemRep output. There are three types of lines: text, entity and relation. The first six fields are used by all three types of lines. The next fields are specific for each line type

Field number	Description (*example value*)
1	BL (*BioMedLEE*) or SE (*SemRep*)
2	PubMed ID
3	Subsection of abstract (*objective*, *results*)
4	Section of abstract ti(*title*) or ab (*abstract*)
5	Sentence id
6	Line type, one of: '*text*', '*entity*', '*relation*'
7. Text	Sentence text
If line type is 'entity' then next fields	
7. Entity	Entity type (*T047* or *disease*)
8. Entity	CUI
9. Entity	Preferred name
10. Entity	Change term (*increase*)
11. Entity	Degree term (*low*)
12. Entity	Negation (*not*)
13. Entity	MetaMap score
14. Entity	Begin character or phrase position
15. Entity	End character position of matched phrase
If line type is 'relation' then next fields	
7–15. Relation	Argument1 related fields
16. Relation	Name of relation (*treat*, *increase*)
17. Relation	Negation of explicit relation or empty
18. Relation	Begin character or phrase position of relation indicator
19. Relation	End character position of relation indicator
20–28. Relation	Argument2 related fields

relations, and modifiers, and for SemRep is plain text. For an *entity* type of line, there are fields specifying the type of entity, UMLS CUI (Concept Unique Identifier), preferred entity name, change and degree of associated change, location of the entity, MetaMap score, and location of the entity in the actual text (start and end position).

For a line of type *relation*, fields 7–15 describe the first argument of the relation in the same format as *entity* line; subsequent fields describe the semantic relation, including the name of the relation, whether it is negated or not, and the start and end positions of the relation in the text. Finally, the second argument of the relation is described in the same way as the first argument in fields 20–28. The specification of the arguments of the relations is currently redundant for ease of experimentation. At a subsequent stage the entities and relations will be associated with identifiers and then arguments of the relations will just be identifiers.

Some of the fields in the common format are specific for only one system, in which case the other system leaves these fields empty. Sometimes the two systems fill a particular field in a different way or format. For example, SemRep uses UMLS semantic types as entity type and BioMedLee uses its own types. BioMedLee identifies the part of the actual text as a phrase identifier within a tagged text format while SemRep uses start and end character positions within a plain text string.

Some of the results presented here were obtained by directly processing the common output format by Unix shell scripts and Perl scripts. Some of the results were produced using SQL statements after the common format output generated by BioMedLee and SemRep was postprocessed with Perl scripts and loaded into a relational database management system.

4 Results

In this section we first replicate Swanson's Raynaud's discovery using the *Maybe_Treats1* discovery pattern. Then we present two hypothetically new therapeutic approaches: one for Huntington disease, based on the *Maybe_Treats1* discovery pattern and one for Parkinson's disease, based on *Maybe_Treats2*. Although we have not done a formal evaluation of our approach, at the end of this section we show evaluation results for the two important components of our methodology, BioMedLee and SemRep.

4.1 Rediscovering Fish Oil for Raynaud's Disease

To illustrate the *Maybe_Treats1* discovery pattern, we show how Swanson's Raynaud's discovery [1] could be replicated. This example also illustrates integration of semantic relation extraction with an existing (co-occurrence based) LBD system. We used the BITOLA [3, 4] LBD system (available at http://www.mf.uni-lj.si/bitola/) and searched for Raynaud's as the starting concept X. Then, among the related concepts Y limited to the semantic group *Physiology*, we found *Blood Viscosity* in the eighth place and *Platelet Aggregation* in the seventeenth place out of 230 concepts from the *Physiology* group that co-occur with Raynaud's. We then submitted the citations in which Raynaud's co-occurs with either *Blood Viscosity* or *Platelet Aggregation* to BioMedLee, which produced five relations in which Raynaud's was associated with an increase in blood viscosity (examples 3 and 4 in Table 2) and one in which Raynaud's was associated with platelet aggregation.

In the next step we used BITOLA to search for concepts co-occurring with blood viscosity or platelet aggregation. Among others, we found *Eicosapentaenoic acid*, which can be found in large quantities in fish oil. After processing the relevant Medline citations with BioMedLee, we obtained several relations in which eicosapentainoic acid was associated with a reduction in blood viscosity (examples 5 and 6 in Table 2). By combining examples 3 and 4 with 5 and 6 we can conclude that eicosapentainoic acid (Z) (and consequently food rich it this acid such as fish oil) might be used to treat Raynaud's (X) because blood viscosity (Y) is increased in Raynaud's and eicosapentainoic acid reduces blood viscosity.

Table 2 Examples of extracted relations by BioMedLee (BL) or SemRep (SR). The relation *Associated_with* shown in column 3, represented a shortened form of *Associated_with_change*

Number	System	Extracted relations	Sentence (or fragment)
1	BL	Associated_with (oxidative stress, iron, increase)	Reducing the oxidative stress associated with increased iron levels
2	SR	Treats(coenzyme Q10,Huntington Disease)	Oral administration of CoQ10 significantly decreased elevated lactate levels in patients with Huntington's disease
3	BL	Associated_with (Raynaud's, blood viscosity, increase)	Local increase of blood viscosity during cold-induced Raynaud's phenomenon
4	BL	Associated_with (Raynaud's, viscosity, increase)	Increased viscosity might be a causal factor in secondary forms of Raynaud's disease, ...
5	BL	Associated_with (eicosapentaenoic acid, blood viscosity, decrease)	We recently reported that eicosapentaenoic acid (EPA) also reduces whole blood viscosity
6	BL	Associated_with (eicosapentaenoic acid, blood viscosity, decrease)	A statistically significant reduction in whole blood viscosity was observed at seven weeks in those patients receiving the eicosapentaenoic acid rich oil
7	BL	Associated_with (Huntington's disease, insulin, decrease)	Huntington's disease transgenic mice develop an age-dependent reduction of insulin mRNA expression and diminished expression of key regulators of insulin gene transcription, ...

4.2 Insulin for Huntington Disease

To illustrate the *Maybe_Treats2* form of the *Maybe_Treats* discovery pattern, we selected *Huntington disease* as a test case. Huntington disease (HD) is an autosomal-dominant inherited neurodegenerative disorder that is characterized by the insidious progressive development of mood disturbances, behavioral changes, involuntary choreiform movements and cognitive impairments. Onset is most common in adulthood, with a typical duration of 15–20 years before premature death. No successful treatment is currently available. We constructed the set of all 5,511 Medline citations (in January, 2006) in which Huntington Disease occurs as a MeSH heading. We first submitted this set to SemRep, which extracted 30,103 relations, out of which 2,139 were *Treats* relations. Of these, 740 Treats relations contained Huntington disease as an argument. These represent current treatments for Huntington (example 2 in Table 2).

Our strategy then was to find relations between HD and changes in substances or body functions which could be potential therapeutic targets for HD. For this we submitted the Huntington citations to BioMedLee, which extracted 18,360 relations, of which 1,912 contained a change, 310 of which were associated with Huntington disease. From the 310 relations, a clinician who is an expert in HD, selected 35 interesting concepts representing neurotransmitters, their receptors or other biologic substances changed in HD. The next step was to find diseases in which these concepts were changed in the same way as in HD. We than assumed that drugs and treatments which are successfully used to treat diseases associated with the same changes in substances and body functions as in HD would be potential new treatments for HD.

By using this approach we discovered an interesting potential new treatment for HD – insulin, which was one of the substances found to be *decreased* in HD (example 7 in Table 2). Although insulin has been attempted for immediate relief of one of the symptoms (chorea) of HD [25], we have not found research on insulin as a general treatment for this disease.

It is known that HD patients develop diabetes mellitus about seven times more often than matched healthy control individuals [26]. The reason for this is unclear, although inappropriate insulin secretion is a potential reason. The transgenic HD mouse model also develops an age-dependent reduction of insulin mRNA expression and diminished expression of key regulators of insulin gene transcription [27].

Strong evidence from studies in humans and animal models suggests the involvement of energy metabolism defects, which may contribute to excitotoxic processes, oxidative damage, and altered gene regulation in the pathogenetic mechanism of HD. Reduced glucose metabolism in affected brain areas of HD patients is a well documented fact used for diagnostic purposes.

We then searched for diseases other than HD with reduced levels of insulin. Expectedly the system identified diabetes mellitus. We thus concluded that insulin treatment, used for diabetes mellitus, might be an interesting drug for HD. Insulin might improve glucose metabolism in the brains of HD patients and thus slow down the pathogenetic process.

4.3 Gabapentin for Parkinson's Disease

This example illustrating the *Maybe_Treats1* pattern for Parkinson's disease uses the same set of articles used for *Maybe_Treats2* above. We selected Parkinson's disease as a starting concept in a modified version of Bitola which integrates co-occurrence based association rules with semantic relations extracted by BioMedLee and SemRep. This version of Bitola is in early development phase and is not yet publicly available.

In order to find potential therapies for the disease, our discovery strategy was first to identify Y concepts (Neuroreactive Substance or Biogenic Amine or Biologically Active Substance), characterized by a "decrease" of some substance in Parkinson's disease and in the second step to find all Z concepts (pharmacological substances) with the opposite change. We limited Y concepts by "change" and got five different concepts. Two of them, *levodopa* and *dopamine* are the mainstream of therapy for decades. The next two of the concepts, *Hommovanilic acid* and *Substance P*, were not selected due to inappropriate context of the relations. A relevant relation was identified in the following sentence: "Postmortem brain studies indicate that patients with *Parkinson's disease* have *decreased* basal ganglia *gamma-aminobutyric acid* function in addition to profound striatal dopamine deficiencies." for *gamma-aminobutyric acid (GABA)*.

In the second step we searched for all Z concepts (pharmacological substance) characterized by an "opposite change". Six substances, all antiepileptics, were identified which were related to GABA in an appropriate way: *gabapentin, Vigabatrin, Tiagabine* and *Topiramate, methamphetamine* and *milacemide* through the following sentences: "*Gabapentin*, probably through the activation of glutamic acid decarboxylase, *leads to* the *increase* in synaptic *GABA*", "*GVG (Vigabatrin) caused* a significant *increase* in *GABA* release, even at concentrations as low as 25 μM", "*Tiagabine* is an antiepileptic drug, which *increases GABA* via selective blockade of GABA reuptake", "*Topiramate increased* brain *GABA*, homocarnosine, and pyrrolidinone to levels that could contribute to its potent antiepileptic action in patients with complex partial seizures." "These results support the hypothesis that long-term administration of *methamphetamine increases* the activity of the striatonigral *GABA* system and thereby reduces the sensitivity of postsynaptic GABA receptors in the SNR." And "The results show that *milacemide increases* the *GABA* content in the GABA pool which is associated with the striatonigral neurons."

GABA is ubiquitous in the nervous system and regarded widely as the principal inhibitory neurotransmitter of the brain. It is also considered as one of the principal vehicles for inhibition in Parkinson's disease. Furthermore, production of inhibitory transmitter GABA in the subthalamic nucleus (STN), suppressing the hyperactive STN, is considered as one of the strategies for gene therapy in the treatment of Parkinson's disease.

In this way we identified selected antiepileptics as a possible therapy for Parkinson's disease. Indeed, some potential benefit of Gabapentin and Toprimate in treatment of Parkinson's disease has been already mentioned in the literature [32, 33].

4.4 Evaluation of BioMedLee and SemRep

Although BioMedLEE has not yet been evaluation for use in LBD, it has been evaluated for two different applications. In Lussier [14], BioMedLEE was combined with a phenotypic ontological organizing system, PhenOS, to create a new system called PhenoGO. PhenoGO associates contextual information with GOA annotations [28] by adding phenotypic information to the protein and GO pairs specified in GOA. The overall PhenoGO system was evaluated for extracting and coding anatomical and cellular information associated with the pairs and for assigning the code to the correct pairs. The results of the evaluation demonstrated that PhenoGO has a precision of 91% and a recall of 92%. Although the results have been computed for the entire PhenoGO system and not for BioMedLEE separately, the high performance of PhenoGO is an indicator of the performance of BioMedLEE because the relations among the genes, GO terms, and phenotypes were determined based on BioMedLEE.

In Borlawsky [29], BioMedLEE was used for a clinical application geared to facilitating clinical practice using Evidence-Based Medicine (EBM). This involved extracting and coding disease, therapy, and drug concepts and their relations from textual sections of Cochrane Reviews, the best standard for obtaining evidence-based medicine. Although BioMedLEE was designed for capturing phenotypic and genotypic relations and not designed for clinical applications or processing of Cochrane Reviews, the study showed that the pertinent information could be extracted and correlated with an overall recall of 80.3% and precision of 75.2%. The most frequent cause of error was due to differences in the semantic classification assigned by BioMedLEE and by the expert, who manually coded the information. For example, the expert manually parsed 'hearing loss' as a problem, but the NLP engine alternatively parsed the phrase as a compositional phrase consisting of a process *hearing* with a change modifier *loss*, which is also correct. Thus, it is likely that performance can be increased by expanding the guidelines to permit certain variations in semantic categorization between the expert and system and by refining the system specifically for the clinical domain, which is not as broad as the complete biomedical domain.

The effectiveness of SemRep in extracting semantic predications from biomedical text has been evaluated in several contexts [24, 30, 31]. In two of these [30, 31], accuracy was assessed after the predications had been subjected to an automatic summarization algorithm. In [30], 306 predications (for predicates ISA, CAUSES, CO-OCCURS_WITH, LOCATION_OF, OCCURS_IN, TREATS) extracted from 1,200 Medline citations were evaluated. Of these, 203 predications were determined to be correct (66% precision). In [eval2], for predicates AFFECTS, CAUSES, COMPLICATES, DISRUPTS, INTERACTS_WITH, ISA, PREVENTS, and TREATS, 148 of 189 predications extracted from 130 Medline citations were judged as correct (78% precision). SemRep was tested for both recall and precision in [24], using a gold standard of 300 sentences randomly generated from 36,577 sentences drawn from a set of Medline citations containing drug and gene co-occurrences. In addition to the predicates addressed in the first two evaluations, predications having such predicates as INHIBITS, STIMULATES, and DISRUPTS were also assessed. SemRep extracted 623 predications from the 300 sentences in the test collection. Of these, 455 were true positives, 168 were false positives, and 375 were false negatives, reflecting recall of 55% (95% confidence interval 49–61%) and precision of 73% (95% confidence interval 65–81%).

5 Discussion and Further Work

Although there are clear advantages in using semantic relation extraction for LBD, there are also some issues that have to be addressed. One is scalability. Ideally all of Medline needs to be processed to support the system we propose. The other issue is accuracy in semantic relation extraction. We presented some general performance evaluation of semantic relation extraction, but in further work we plan to evaluate specifically the extraction of *Associated_with_change* and *Treats*, which are the most important relations in our method. We also plan to evaluate the performance of the overall LDB method. Because of these issues, we believe that for the near future, the best approach would be the integration of semantic relation extraction with co-occurrence-based LBD. In further work we plan to better integrate the BITOLA LBD system with SemRep and BioMedLee. Currently, the user has to run the three systems separately and the output is combined with various scripts in a way which is not very user-friendly.

Another research contribution is the use of two natural language processing systems, namely SemRep and BioMedLee, to extract the kind of relations they are best at capturing. This entailed developing a common format for each system's output. To our knowledge this is the first time two different natural language processing systems have been utilized together to capture different types of semantic relations. We plan to combine BioMedLee's change detection with SemRep's relations in order to obtain a larger number of binary relations with a change. Namely, SemRep may find a binary relation whereas BioMedLEE may not, but BioMedLEE may have found a change in one of the arguments of the relation that SemRep found. Currently we have a large number of unary change relations which are not associated directly with another concept. Another way to improve the extraction of change relations is by analyzing the cases in which the change was not captured and creating better extraction rules.

Yet another research contribution is the notion of a *discovery pattern* which is based on semantic relations and allows more precise hypothesis generation. Here we have presented one such pattern, *Maybe_Treats*, but we plan to develop other discovery patterns as well.

We plan to develop a user-friendly web-based interface which will allow public access to our methodology. It should allow among other things ranking of potentially new discoveries based on a heuristic ranking procedure not yet developed.

6 Conclusions

Literature-based discovery (LBD) is a method for automatically generating hypotheses from the research literature. Currently LBD systems depend exclusively on co-occurrence based methods for finding relations between concepts. We presented a new method aimed at improving LBD. It is based on semantic predications, which are extracted from text using the combined results of two natural language processing systems. Additionally, the change associated with the arguments of the predications, is also extracted. We also introduced the notion of a *discovery pattern*. The proposed system has the potential to produce a smaller number of false positive discoveries while, at the same time, facilitating user evaluation and review of potentially new relations. Finally, it can support explanation of the discovery produced.

Using our methodology we successfully replicated Swanson's Raynaud's – fish oil discovery. Furthermore, we generated some interesting potentially new therapeutic approaches for Huntington disease and for Parkinson's disease.

We believe that the future of literature-based discovery lies in developing specific discovery patterns for particular discovery tasks based on semantic relations further integrated with co-occurrence-based approaches.

Acknowledgements The part of this research done at Columbia University was supported by grants LM007659 and LM008635 from the National Institutes of Health. This study was supported in part by the Intramural Research Programs of the National Institutes of Health, National Library of Medicine.

References

1. Swanson, D.R.: Fish oil, Raynaud's syndrome, and undiscovered public knowledge. Perspect Biol Med **30** (1986) 7–18
2. Swanson, D.R., Smalheiser, N.R.: An interactive system for finding complementary literatures: a stimulus to scientific discovery. Artif Intell **91** (1997) 183–203
3. Hristovski, D., Peterlin, B., Mitchell, J.A., Humphrey, S.M.: Using literature-based discovery to identify disease candidate genes. Int J Med Inform **74** (2005) 289–298
4. Hristovski, D., Stare, J., Peterlin, B., Dzeroski, S.: Supporting discovery in medicine by association rule mining in Medline and UMLS. Medinfo **10** (2001) 1344–1348
5. Weeber, M., Klein, H., Aronson, A.R., Mork, J.G., de Jong-van den Berg, L.T., Vos, R.: Text-based discovery in biomedicine: the architecture of the DAD-system. Proc AMIA Symp (2000) 903–907
6. Gordon, M.D., Lindsay, R.K.: Toward discovery support systems: a replication, re-examination, and extension of Swanson's work on literature-based discovery of a connection between Raynaud's and fish oil. J Am Soc Inf Sci **47** (1996) 116–128
7. Gordon, M.D., Dumais, S.: Using latent semantic indexing for literature based discovery. J Am Soc Inf Sci **49** (1998) 674–685
8. Wren, J.D.: Extending the mutual information measure to rank inferred literature relationships. BMC Bioinformatics **5** (2004) 145
9. Pratt, W., Yetisgen-Yildiz, M.: LitLinker: capturing connections across the biomedical literature. In Proceedings of the 2nd International Conference on Knowledge Capture. ACM Press, Sanibel Island, FL, USA (2003)
10. Fuller, S.S., Revere, D., Bugni, P.F., Martin, G.M.: A knowledgebase system to enhance scientific discovery: Telemakus. Biomed Digit Libr **1** (2004) 2
11. Hu, X.: Mining novel connections from large online digital library using biomedical ontologies. Libr Manage **26** (2005) 261–270
12. Srinivasan, P., Libbus, B.: Mining MEDLINE for implicit links between dietary substances and diseases. Bioinformatics **20 Suppl 1** (2004) I290–I296
13. Rindflesch, T.C., Fiszman, M.: The interaction of domain knowledge and linguistic structure in natural language processing: interpreting hypernymic propositions in biomedical text. J Biomed Inform **36** (2003) 462–477
14. Lussier, Y., Borlawsky, T., Rappaport, D., Liu, Y., Friedman, C.: PhenoGO: assigning phenotypic context to gene ontology annotations with natural language processing. Pac Symp Biocomput (2006). pp. 64–75
15. Weeber, M., Kors, J.A., Mons, B.: Online tools to support literature-based discovery in the life sciences. Brief Bioinform **6** (2005) 277–286

16. Aronson, A.R.: Effective mapping of biomedical text to the UMLS Metathesaurus: the MetaMap program. Proc AMIA Symp (2001) 17–21
17. Agrawal, R., Mannila, H., Srikant, R., Toivonen, H., Verkamo, A.I.: Fast discovery of association rules. In: Fayyad, U. (ed.): Advances in Knowledge Discovery and Data mining. MIT Press, Cambridge, MA (1996), pp. 307–328
18. Friedman, C., Alderson, P.O., Austin, J.H., Cimino, J.J., Johnson, S.B.: A general natural-language text processor for clinical radiology. J Am Med Inform Assoc **1** (1994) 161–174
19. Friedman, C., Shagina, L., Lussier, Y., Hripcsak, G.: Automated encoding of clinical documents based on natural language processing. J Am Med Inform Assoc **11** (2004) 392–402
20. Humphreys, B.L., Lindberg, D.A., Schoolman, H.M., Barnett, G.O.: The Unified Medical Language System: an informatics research collaboration. J Am Med Inform Assoc **5** (1998) 1–11
21. Friedman, C., Borlawsky, T., Shagina, L., Xing, H.R., Lussier, Y.A.: Bio-ontology and text: bridging the modeling gap. Bioinformatics **22** (2006) 2421–2429
22. McCray, A.T., Srinivasan, S., Browne, A.C.: Lexical methods for managing variation in biomedical terminologies. Proc Annu Symp Comput Appl Med Care (1994) 235–239
23. Smith, L., Rindflesch, T., Wilbur, W.J.: MedPost: a part-of-speech tagger for biomedical text. Bioinformatics **20** (2004) 2320–2321
24. Ahlers, C., Fiszman, M., Demner-Fushman, D., Lang, F.-M., Thomas, C.R.: Extracting semantic predications from Medline citations for pharmacogenomics. Pac Symp Biocomput (2007) 209–220
25. Quinn, N.P., Lang, A.E., Marsden, C.D.: Insulin-induced hypoglycaemia does not abolish chorea. J Neurol Neurosurg Psychiatry **45** (1982) 1169–1170
26. Ristow, M.: Neurodegenerative disorders associated with diabetes mellitus. J Mol Med **82** (2004) 510–529
27. Andreassen, O.A., Dedeoglu, A., Stanojevic, V., Hughes, D.B., Browne, S.E., Leech, C.A., Ferrante, R.J., Habener, J.F., Beal, M.F., Thomas, M.K.: Huntington's disease of the endocrine pancreas: insulin deficiency and diabetes mellitus due to impaired insulin gene expression. Neurobiol Dis **11** (2002) 410–424
28. Camon, E., Magrane, M., Barrell, D., Lee, V., Dimmer, E., Maslen, J., Binns, D., Harte, N., Lopez, R., Apweiler, R.: The Gene Ontology Annotation (GOA) Database: sharing knowledge in Uniprot with Gene Ontology. Nucleic Acids Res **32** (2004) D262–D266
29. Borlawsky, T., Friedman, C., Lussier, Y.: Generating executable knowledge for evidence-based medicine using natural language and semantic processing. AMIA Annu Symp Proc (2006)
30. Fiszman, M., Rindflesch, T.C., Kilicoglu, H.: Abstraction summarization for managing the biomedical research literature. Proc HLTNAACL Workshop on Computational Lexical Semantics (2004) 76–83
31. Fiszman, M., Rindflesch, T., Kilicoglu, H.: Summarizing drug information in Medline citations. Proc AMIA Annu Symp (2006)
32. Van Blercom, N., Lasa, A., Verger, K., Masramón, X., Sastre, V.M., Linazasoro, G: Effects of gabapentin on the motor response to levodopa: a double-blind, placebo-controlled, crossover study in patients with complicated Parkinson disease. Clin Neuropharmacol **27** (2004) 124–128
33. Silverdale, M.A., Nicholson, S.L., Crossman, A.R., Brotchie, J.M.: Topiramate reduces levodopa-induced dyskinesia in the MPTP-lesioned marmoset model of Parkinson's disease. Mov Disord **20** (2005) 403–409

Information Retrieval in Literature-Based Discovery

W. Hersh

Abstract Finding and accessing relevant information is essential for wider use of literature-based discovery (LBD). This chapter provides an overview of information retrieval (IR) with a focus on its role in LBD. It covers the major approaches to indexing and retrieval, followed by a description of research evaluating them. The chapter concludes with an overview of IR techniques used for LBD and promising directions for the future.

1 Introduction

Information retrieval (IR) is the field concerned with the indexing and retrieval of knowledge-based information [25]. Although the name implies the retrieval of any type of information, the field has traditionally focused on retrieval of text-based documents, reflecting the type of information that was initially available by this early application of computer use. However, with the growth of multimedia content, including images, video, and other types of information, IR has broadened considerably. The proliferation of IR systems and on-line content have also changed the notion of the libraries, which have traditionally been viewed as buildings or organizations. However, the development of the Internet and new models for publishing have challenged this notion as well, and new *digital libraries* have emerged [6].

A perspective of the role of IR is provided in Fig. 1, which shows the flow of extracting knowledge from the scientific literature. IR typically focuses on the initially narrowing of the broad literature, ideally passing off a more focused set of articles for the more intensive processing required for extracting and structuring knowledge.

W. Hersh
Department of Medical Informatics and Clinical Epidemiology,
School of Medicine, Oregon Health and Science University,
3181 SW Sam Jackson Park Rd., Portland, OR 97239, USA
hersh@ohsu.edu

P. Bruza and M. Weeber (eds.), *Literature-based Discovery,*
Springer Series in Information Science and Knowledge Management 15.

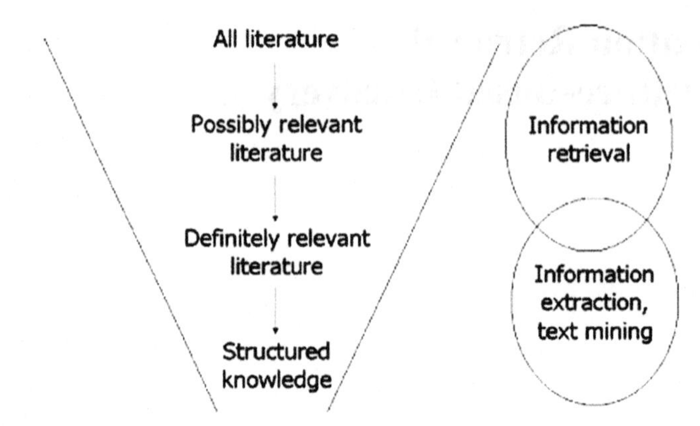

Fig. 1 Information retrieval in context

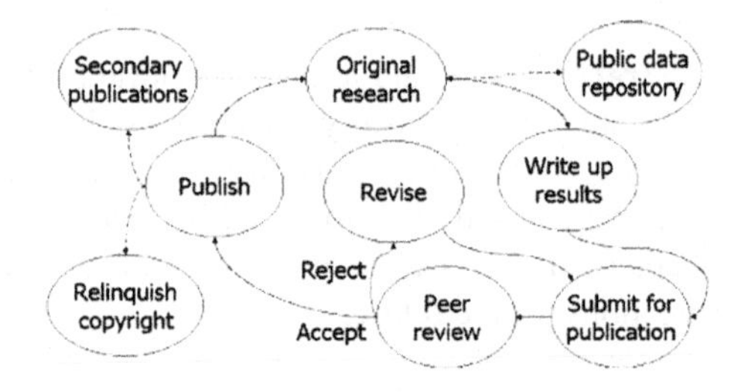

Fig. 2 The "life cycle" of scientific information

IR systems and digital libraries store and disseminate knowledge-based information, the type of information that is derived and organized from observational and experimental research. Knowledge-based information is most commonly provided in scientific journals and proceedings but can be published in a wide variety of other forms, including books, Web sites, and so forth. Figure 2 depicts the "life cycle" of primary literature, which is derived from original research and whose publication is dependent upon the peer review process that insures the methods, results, and interpretation of results meets muster with one's scientific peers. In some fields, such as genomics, there is an increasing push for original data to enter public repositories. In most fields, primary information is summarized in secondary publications, such as review articles and textbooks. Also in most fields, the authors relinquish the copyright of their papers to publishers, although there is increasing resistance to this, as described later in this chapter.

IR systems have usually, although not always, been applied to knowledge-based information, which can be subdivided in other ways. *Primary knowledge-based information* (also called primary literature) is original research that appears in journals, books, reports, and other sources. This type of information reports the initial discovery of health knowledge, usually with either original data or re-analysis of data (e.g., systematic reviews and/or meta-analyses).

Secondary knowledge-based information consists of the writing that reviews, condenses, and/or synthesizes the primary literature. As seen in Fig. 1, secondary literature emanates from original publications. The most common examples of this

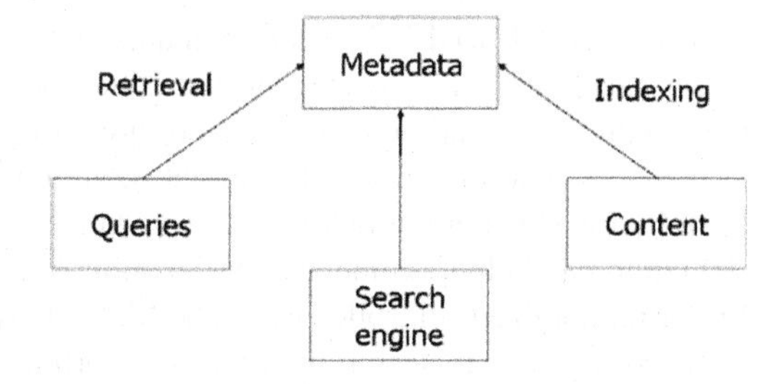

Fig. 3 The information retrieval process [25]

type of literature are books, monographs, and review articles in journals and other publications. Secondary literature also includes opinion-based writing such as editorials and position or policy papers.

Now that we have had a general overview of knowledge-based information, we can look in further detail at IR systems. A model for the IR system and the user interacting with it is shown in Fig. 3 [25]. The ultimate goal of a user of an IR system is to access content, which may be in the form of a digital library. In order for that content to be accessible, it must be described with metadata. The major intellectual processes of IR are *indexing* and *retrieval.* In the following sections, we will discuss content, indexing, and retrieval, followed by an overview of how IR systems are evaluated and issues concerning digital libraries.

2 Content

The ultimate goal of IR systems and digital libraries is to deliver information to users for specific tasks. It is useful to classify the different types of knowledge-based information to better understand the issues in its indexing and retrieval. In this section, we classify content into bibliographic, full-text, and more structured databases/collections.

2.1 Bibliographic

The first category consists of *bibliographic content.* It includes what was for decades the mainstay of IR systems: literature reference databases. Also called *bibliographic databases*, this content consists of citations or pointers to the scientific literature (i.e., journal articles). An example of such a bibliographic database is MEDLINE, which is produced by the National Library of Medicine (NLM) and contains bibliographic references to the articles, editorials, and letters to the editors in approximately 4,500 biomedical journals. At present, about 500,000 references are added to MEDLINE yearly. It now contains over 16 million references. MEDLINE is licensed to a variety of information providers and also available directly for free on the NLM Web site in the Pubmed system (pubmed.gov).

The current MEDLINE record contains up to 49 fields. Probably the most commonly used fields are the title, abstract, and indexing terms. But other fields contain specific information that may be of great importance to smaller audiences. For example, a genomics researcher might be highly interested in the Supplementary Information (SI) field to link to genomic databases. Likewise, the Publication Type (PT) field can help to clinicians, designating whether an article is a practice guideline or randomized controlled trial. The NLM also partitions MEDLINE into subsets for users wishing to search on a focused portion of the database, such as *AIDS* or *Complementary and Alternative Medicine*.

There are, of course, plenty of non-medical bibliographic databases as well. The *ACM Guide to Computing Literature* is part of the *ACM Portal* for computer science literature, which also includes the *ACM Digital Library* that contains the full text of articles published by ACM Press. Another source of on-line bibliographic information is *Google Scholar*, which lacks the richness of more focused bibliographic databases but makes up for it in breadth of coverage.

A second, more modern type of bibliographic content is the Web catalog. There are increasing numbers of such catalogs, which consist of Web pages containing mainly links to other Web pages and sites. Some well-known Web catalogs include Yahoo (www.yahoo.com) and Open Directory (dmoz.org). There are also many Web catalogs specific to certain subject domains, Some well-known medical Web catalogs, for example, include:

- INTUTE Health and Life Sciences (http://www.intute.ac.uk/healthandlifesciences/) – catalog of high-quality health information maintained in UK
- MedlinePLUS (medlineplus.gov) – catalog of consumer-oriented medical information maintained by the NLM [35]
- HealthFinder (healthfinder.gov) – consumer-oriented health information maintained by the Office of Disease Prevention and Health Promotion of the US Department of Health and Human Services

2.2 Full-Text

The second type of content is *full-text content*. A large component of this content consists of the online versions of books and periodicals. A wide variety of the traditional paper-based biomedical literature, from textbooks to journals, is now available electronically. The electronic versions may be enhanced by measures ranging from the provision of supplemental data in a journal article to linkages and multimedia content in a textbook. The final component of this category is the Web site. Admittedly the diversity of information on Web sites is enormous, and sites may include every other type of content described in this chapter. However, in the context of this category, "Web site" refers to a localized collection (that may be large) of static and dynamic pages at a discrete Web location.

Most scientific journals are now published in electronic form. Electronic publication not only allows easier access, but additional features not possible in print versions. For example, journal Web sites can provide additional data with additional figures and tables, results, images, and even raw data. A journal Web site also allows more dialogue about articles than could be published in a Letters to the Editor section of a print journal. Electronic publication also allows true bibliographic linkages, both to other full-text articles and to the bibliographic record. The Web also allows linkage directly from bibliographic databases to full text. In fact, some bibliographic databases such MEDLINE database now have a field for the Web address of the full-text paper.

Electronic full-text journals may be produced by the original publisher (e.g., Springer, ACM, etc.) or by a value-added publisher. An example of the latter is Highwire Press (www.highwire.org), which works with publishers to produce electronic versions of their journals. The Highwire system provides a retrieval interface that searches over the complete online contents for a given journal. Users can search for authors, words limited to the title and abstract, words in the entire article, and within a date range. The interface also allows searching by citation by entering volume number and page as well as searching over the entire collection of journals that use Highwire. Users can also browse through specific issues as well as collected resources.

The most common full-text secondary literature source is the traditional textbook, an increasing number of which are available in electronic form. A common approach with textbooks is to bundle multiple books, sometimes with linkages across them. Electronic textbooks offer additional features beyond text from the print version. While many print textbooks do feature high-quality images, electronic versions offer the ability to have more pictures and illustrations. They also have the ability to use sound and video, although few do at this time. As with full-text journals, electronic textbooks can link to other resources, including journal references and the full articles. Many Web-based textbook sites also provide access to continuing education self-assessment questions and medical news. In addition, electronic textbooks let authors and publishers provide more frequent updates of the information than is allowed by the usual cycle of print editions, where new versions come out only every 2–5 years.

As noted above, Web sites are another form of full-text information. One of the largest users of the Web to provide information on a variety of topics is the U.S. government. Examples include:

- The THOMAS system of current and pending legislation (thomas.loc.gov)
- The US Consumer Product Safety Commission (www.cpsc.gov)
- Travel information for various parts of the world (http://www.cdc.gov/travel/)

2.3 Databases/Collections

The final category consists of *databases and other specific collections* of content. These resources are usually not stored as freestanding Web pages but instead are often housed in database management systems. Where full-text content ends and this category begins is admittedly somewhat fuzzy, but clearly there is a difference from an IR standpoint between pure text and more multimedia-rich sources of information. This content can be further subcategorized into discrete information types:

- Multimedia databases – collections of images, videos, sounds, etc.
- Citation databases – bibliographic linkages of scientific literature
- Other domain-specific databases

A great number of image databases are available on the Web, from historical archives to image collections from focused areas, such as airliners (airlines.net) and the US National Aeronautics and Space Administration (NASA, http://images.jsc.nasa.gov/). The Web is full of biomedical image collections as well. One well-known collection is the Visible Human Project of the NLM, which consists of three-dimensional representations of normal male and female bodies [42]. This resource is built from cross-sectional slices of cadavers, with sections of 1 mm in the male and 0.3 mm in the female. Also available from each cadaver are transverse computerized tomography (CT) and magnetic resonance (MR) images. In addition to the images themselves, a variety of searching and browsing interfaces have been created which can be accessed via the project Web site (http://www.nlm.nih.gov/research/visible/visible_human.html).

Citation databases provide linkages to articles that cite others across the scientific literature. The best-known citation databases are the *Science Citation Index* (SCI, ISI Thompson) and *Social Science Citation Index* (SSCI, ISI Thompson). A recent development is the *Web of Science*, a Web-based interface to these databases. Another system for citation indexing is the *Research Index* (formerly called *CiteSeer*, citeseer.nj.nec.com) [32]. This index uses a process called *autonomous citation indexing* that adds citations into its database by automatically processing of papers from the Web. It also attempts to identify the context of citations, showing words similar across citations such that the commonality of citing papers can be observed.

One example category of domain-specific databases is the *model organism database* [37]. This kind of database brings together bibliographic, full text, and other databases of sequences, structure, and function for organisms such as the mouse [9] and *Saccharomyces* yeast [3]. Another well-known aggregation of genomics information is the SOURCE (source.stanford.edu) database, which aggregates information from many other sources about individuals genes in species [14].

3 Indexing

Most modern commercial IR systems index their content in two ways. In *manual indexing*, human indexers, usually using standardized terminology, assign indexing terms and attributes to documents, often following a specific protocol. Manual indexing is typically done using *controlled vocabularies*, which consist of the set of allowable terms and relationships between them. In *automated indexing*, on the other hand, computers make the indexing assignments, usually limited to breaking out each word in the document (or part of the document) as an indexing term.

Manual indexing is used most commonly with bibliographic databases. In this age of proliferating electronic content, such as online textbooks, practice guidelines, and multimedia collections, manual indexing has become either too expensive or outright unfeasible for the quantity and diversity of material now available. Thus there are increasing numbers of databases that are indexed only by automated means.

3.1 Controlled Vocabularies

Before discussing specific vocabularies, it is useful to define some terms, since different writers attach different definitions to the various components of thesauri. A *concept* is an idea or object that occurs in the world, such as the condition under which human blood pressure is elevated. A *term* is the actual string of one or more words that represent a concept, such as *automobile* or *car*. One of these string forms is the *preferred* or *canonical* form, such as *automobile* in the present example. When one or more terms can represent a concept, the different terms are called *synonyms*.

A controlled vocabulary usually contains a list of terms that are the canonical representations of the concepts. They are also called *thesauri* and contain relationships between terms, which typically fall into three categories:

1. Hierarchical – terms that are broader or narrower. The hierarchical organization not only provides an overview of the structure of a thesaurus but also can be used to enhance searching.
2. Synonymous – terms that are synonyms, allowing the indexer or searcher to express a concept in different words.
3. Related – terms that are not synonymous or hierarchical but are somehow otherwise related. These usually remind the searcher of different but related terms that may enhance a search.

One well-known thesaurus used for indexing bibliographic databases is the Medical Subject Headings (MeSH) vocabulary is used by the NLM for MEDLINE and other databases [13]. MeSH contains over 21,000 *subject headings* (the word MeSH uses to denote the canonical representation of its concepts). It also contains over 100,000 supplementary concept records in a separate chemical thesaurus. In addition, MeSH contains the three types of relationships described in the previous paragraph:

1. Hierarchical – MeSH is organized hierarchically into 15 trees, such as *Diseases*, *Organisms*, and *Chemicals and Drugs*.
2. Synonymous – MeSH contains a vast number of entry terms, which are synonyms of the headings.
3. Related – terms that may be useful for searchers to add to their searches when appropriate are suggested for many headings.

The MeSH vocabulary files, their associated data, and their supporting documentation are available on the NLM's MeSH Web site (www.nlm.nih.gov/mesh/). There is also a browser that facilitates exploration of the vocabulary (www.nlm.nih.gov/mesh/MBrowser.html).

There are features of MeSH designed to assist indexers in making documents more retrievable. One of these is subheadings, which are qualifiers of subject headings that narrow the focus of a term. In the *Hypertension*, for example, the focus of an article may be on the diagnosis, epidemiology, or treatment of the condition. Another feature of MeSH that helps retrieval is check tags. These are MeSH terms that represent certain facets of medical studies, such as age, gender, human or nonhuman, and type of grant support. Related to check tags are the geographical locations in the Z tree. Indexers must also include these, like check tags, since the location of a study (e.g., *Oregon*) must be indicated. Another feature gaining increasing importance for EBM and other purposes is the publication type, which describes the type of publication or the type of study. A searcher who wants a review of a topic may choose the publication type *Review* or *Review Literature*. Or, to find studies that provide the best evidence for a therapy, the publication type *Meta-Analysis*, *Randomized Controlled Trial*, or *Controlled Clinical Trial* would be used.

3.2 Manual Indexing

Manual indexing of bibliographic content is the most common and developed use of such indexing. Bibliographic manual indexing is usually done by means of a controlled vocabulary of terms and attributes. Most databases utilizing human indexing usually have a detailed protocol for assignment of indexing terms from the thesaurus. The MEDLINE database is a good example. The principles of MEDLINE indexing were laid out in the two-volume MEDLARS Indexing Manual [11, 12]. Subsequent modifications have occurred with changes to MEDLINE, other databases, and MeSH over the years. The major concepts of the article, usually from two to five headings, are designed as central concept headings, and designated in the MEDLINE record by an asterisk. The indexer is also required to assign appropriate subheadings. Finally, the indexer must also assign check tags, geographical locations, and publication types.

Few full-text resources are manually indexed. One type of indexing that commonly takes place with full-text resources, especially in the print world, is that performed for the index at the back of the book. However, this information is rarely used in IR systems; instead, most online textbooks rely on automated indexing (see below).

Manual indexing of Web content is challenging. With several billion pages of content, manual indexing of more than a fraction of it is not feasible. On the other hand, the lack of a coherent index makes searching much more difficult, especially when specific resource types are being sought. A simple form of manual indexing of the Web takes place in the development of the Web catalogs and aggregations as described above. These catalogs make not only explicit indexing about subjects and other attributes, but also implicit indexing about the quality of a given resource by the decision of whether to include it in the catalog.

Two major approaches to manual indexing have emerged on the Web, which are not mutually incompatible. The first approach, that of applying metadata to Web pages and sites, is exemplified by the Dublin Core Metadata Initiative (DCMI, www.dublincore.org). The second approach, to build directories of content, is further described below.

The goal of the DCMI has been to develop a set of standard data elements that creators of Web resources can use to apply metadata to their content [52]. The specification has defined 15 elements, as shown in Table 1. The DCMI has been anointed a standard by the National Information Standards Organization (NISO) with the designation Z39.85.

While Dublin Core Metadata was originally envisioned to be included in HTML Web pages, it became apparent that many non-HTML resources exist on the Web and that there are reasons to store metadata external to Web pages. For example, authors of Web pages might not be the best people to index pages or other entities

Table 1 Elements of Dublin Core metadata

Dublin Core element	Definition
DC.title	The name given to the resource
DC.creator	The person or organization primarily responsible for creating the intellectual content of the resource
DC.subject	The topic of the resource
DC.description	A textual description of the content of the resource
DC.publisher	The entity responsible for making the resource available in its present form
DC.date	A date associated with the creation or availability of the resource
DC.contributor	A person or organization not specified in a creator element who has made a significant intellectual contribution to the resource but whose contribution is secondary to any person or organization specified in a creator element
DC.type	The category of the resource
DC.format	The data format of the resource, used to identify the software and possibly hardware that might be needed to display or operate the resource
DC.identifier	A string or number used to uniquely identify the resource
DC.source	Information about a second resource from which the present resource is derived
DC.language	The language of the intellectual content of the resource
DC.relation	An identifier of a second resource and its relationship to the present resource
DC.coverage	The spatial or temporal characteristics of the intellectual content of the resource
DC.rights	A rights management statement, an identifier that links to a rights management statement, or an identifier that links to a service providing information about rights management for the resource

Source: www.dublincore.org

might wish to add value by their own indexing of content. An emerging standard for cataloging metadata is the *Resource Description Framework* (RDF) [34]. A framework for describing and interchanging metadata, RDF is usually expressed in XML. RDF also forms the basis of what some call the future of the Web as a repository not only of content but also knowledge, which is also referred to as the *Semantic Web* [5]. Dublin Core Metadata (or any type of metadata) can be represented in RDF [4].

Another approach to manually indexing content on the Web has been to create directories of content. The first major effort to create these was the Yahoo! search engine, which created a subject hierarchy and assigned Web sites to elements within it (www.yahoo.com). When concern began to emerge that the Yahoo directory was proprietary and not necessarily representative of the Web community at large [10], an alternative movement emerged, the Open Directory Project.

Manual indexing has a number of limitations, the most significant of which is inconsistency. This has been studied in MEDLINE. Funk and Reid [18] evaluated indexing inconsistency in MEDLINE by identifying 760 articles that had been indexed twice by the NLM. The most consistent indexing occurred with check tags and central concept headings, which were only indexed with a consistency of 61–75%. The least consistent indexing occurred with subheadings, especially those assigned to non-central concept headings, which had a consistency of less than 35%. Manual indexing also takes time. While it may be feasible with the large resources the NLM has to index MEDLINE, it is probably impossible with the growing amount of content on Web sites and in other full-text resources. Indeed, the NLM has recognized the challenge of continuing to have to index the growing body of biomedical literature and is investigating automated and semi-automated means of doing so [2].

3.3 Automated Indexing

In automated indexing, the work is done by a computer. Although the mechanical running of the automated indexing process lacks cognitive input, considerable intellectual effort may have gone into building the automated indexing system. In this section, we will focus on the automated indexing used in operational IR systems, namely the indexing of documents by the words they contain.

Some may not think of extracting all the words in a document as "indexing," but from the standpoint of an IR system, words are descriptors of documents, just like human-assigned indexing terms. Most retrieval systems actually use a hybrid of human and word indexing, in that the human-assigned indexing terms become part of the document, which can then be searched by using the whole controlled vocabulary term or individual words within it. With the development of full-text resources in the 1980s and 1990s, systems that only used word indexing began to emerge. This trend increased with the advent of the Web.

Word indexing is typically done by taking all consecutive alphanumeric characters between white space, which consists of spaces, punctuation, carriage returns, and other nonalphanumeric characters. Systems must take particular care to apply the same process to documents and the user's queries, especially with characters such as hyphens and apostrophes. Some systems go beyond simple identification of words and attempt to assign weights to words that represent their importance in the document [39].

Many systems using word indexing employ processes to remove common words or conflate words to common forms. The former consists of filtering to remove stop words, which are common words that always occur with high frequency and usually of little value in searching. The *stop list*, also called a *negative dictionary*, varies in size in different systems. Examples of stop lists include the 250-word list of van Rijsbergen [49], the 471-word list of Fox [16], and the PubMed stop list [1]. Conflation of words to common forms is done via *stemming*, the purpose of which is to ensure words with plurals and common suffixes (e.g., *-ed, -ing, -er, -al*) are always indexed by their stem form [17]. For example, the words *cough*, *coughs*, and *coughing* are all indexed via their stem *cough*. Stop word removal and stemming also reduce the size of indexing files and lead to more efficient query processing.

A commonly used approach for term weighting is TF*IDF weighting, which combines the inverse document frequency (IDF) and term frequency (TF). The IDF is the logarithm of the ratio of the total number of documents to the number of documents in which the term occurs. It is assigned once for each term in the database, and it correlates inversely with the frequency of the term in the entire database. The usual formula used is

$$IDF(term) = \log \frac{\text{number of documents in database}}{\text{number of documents with term}} + 1. \tag{1}$$

The TF is a measure of the frequency with which a term occurs in a given document and is assigned to each term in each document, with the usual formula:

$$TF(term, document) = \text{frequency of term in document}. \tag{2}$$

In TF*IDF weighting, the two terms are combined to form the indexing weight, WEIGHT:

$$WEIGHT(term, document) = TF(term, document)^{*} IDF(term). \tag{3}$$

Another automated indexing approach generating increased interest is the use of link-based methods, fueled by the success of the Google (www.google.com) search engine. This approach gives weight to pages based on how often they are cited by other pages. The *PageRank* algorithm is mathematically complex, but can be viewed as giving more weight to a Web page based on the number of other pages that link to it, especially when those pages contain a high number of links to them [7, 31]. In general, authoritative pages are likely to have a very high PageRank, whereas more obscure pages tend to have a lower PageRank.

Similar to manual indexing, word-based automated indexing has a number of limitations, including:

- Synonymy – different words may have the same meaning, such as *high* and *elevated.*
- Polysemy – the same word may have different meanings or senses. For example, the word *lead* can refer to an element or to a part of an electrical device.
- Content – words in a document may not reflect its focus. For example, an article describing *automobiles* may make mention in passing of other concepts, such as *airplanes*, which are not the focus of the article.
- Context – words take on meaning based on other words around them. For example, the relatively common words *high, blood,* and *pressure,* take on added meaning when occurring together in the phrase *high blood pressure.*
- Morphology – words can have suffixes that do not change the underlying meaning, such as indicators of plurals, various participles, adjectival forms of nouns, and nominalized forms of adjectives.
- Granularity – queries and documents may describe concepts at different levels of a hierarchy. For example, a user might query for *countries* involved in World War II, but the documents might describe specific countries themselves, such as *Germany, the United Kingdom,* and *the United States.*

4 Retrieval

There are two broad approaches to retrieval. *Exact-match searching* allows the user precise control over the items retrieved. *Partial-match searching*, on the other hand, recognizes the inexact nature of both indexing and retrieval, and instead attempts to return the user content ranked by how close it comes to the user's query. After general explanations of these approaches, we will describe actual systems that access the different types of biomedical content.

4.1 Exact-Match

In exact-match searching, the IR system gives the user all documents that exactly match the criteria specified in the search statement(s). Since the Boolean operators AND, OR, and NOT are usually required to create a manageable set of documents, this type of searching is often called *Boolean searching*. Furthermore, since the user typically builds sets of documents that are manipulated with the Boolean operators, this approach is also called *set-based searching*. Most of the early operational IR systems in the 1950s through 1970s used the exact-match approach, even though Salton was developing the partial-match approach in research systems during that time [41]. In modern times, exact-match searching tends to be associated with retrieval from bibliographic databases, while the partial-match approach tends to be used with full-text searching.

Typically the first step in exact-match retrieval is to select terms to build sets. Other attributes, such as the author name or publication type, may be selected to build sets as well. Once the search term(s) and attribute(s) have been selected, they are combined with the Boolean operators. The Boolean AND operator is typically used to narrow a retrieval set to contain only documents about two or more concepts. The Boolean OR operator is usually used when there is more than one way to express a concept. The Boolean NOT operator is often employed as a subtraction operator that must be applied to another set. Some systems more accurately call this the ANDNOT operator.

Some systems allow terms in searches to be expanded by using the *wild-card character*, which adds all words to the search that begin with the letters up until the wild-card character. This approach is also called *truncation*. Unfortunately there is no standard approach to using wild-card characters, so syntax for them varies from system to system. PubMed, for example, allows a single asterisk at the end of a word to signify a wild-card character. Thus the query word *can** will lead to the words *cancer* and *Candida*, among others, being added to the search. The AltaVista search engine (www.altavista.com) takes a different approach. The asterisk can be used as a wild-card character within or at the end of a word but only after its first three letters. For example, *col*r* will retrieve documents containing *color*, *colour*, and *colder*.

4.2 Partial-Match

Although partial-match searching was conceptualized in the 1960s, it did not see widespread use in IR systems until the advent of Web search engines in the 1990s. This is most likely because exact-match searching tends to be preferred by "power users" whereas partial-match searching is preferred by novice searchers, the ranks of whom have increased substantially with the growth and popularity of the Web. Whereas exact-match searching requires an understanding of Boolean operators and (often) the underlying structure of databases (e.g., the many fields in MEDLINE), partial-match searching allows a user to simply enter a few terms and start retrieving documents.

The development of partial-match searching is usually attributed to Salton [39]. Although partial-match searching does not exclude the use of nonterm attributes of documents, and for that matter does not even exclude the use of Boolean operators (e.g., [40]), the most common use of this type of searching is with a query of a small number of words, also known as a *natural language query*. Because Salton's approach was based on vector mathematics, it is also referred to as the *vector-space model* of IR. In the partial-match approach, documents are typically ranked by their closeness of fit to the query. That is, documents containing more query terms will likely be ranked higher, since those with more query terms will in general be more likely to be relevant to the user. As a result this process is called relevance ranking. The entire approach has also been called *lexical-statistical retrieval*.

The most common approach to document ranking in partial-match searching is to give each a score based on the sum of the weights of terms common to the document and query. Terms in documents typically derive their weight from the TF*IDF calculation described above. Terms in queries are typically given a weight of one if the term is present and zero if it is absent. The following formula can then be used to calculate the document weight across all query terms:

$$Document\ weight = \sum_{all\ query\ terms} Weight\ of\ term\ in\ query^{*}\ Weight\ of\ term\ in\ document. \quad (4)$$

This may be thought of as a giant OR of all query terms, with sorting of the matching documents by weight. The usual approach is for the system to then perform the same stop word removal and stemming of the query that was done in the indexing process. (The equivalent stemming operations must be performed on documents and queries so that complementary word stems will match.)

5 Evaluation

There has been a great deal of research over the years devoted to evaluation of IR systems. As with many areas of research, there is controversy as to which approaches to evaluation best provide results that can assess their searching and the systems they are using. Many frameworks have been developed to put the results in context. One of these frameworks organizes evaluation around six questions that someone advocating the use of IR systems might ask [28]:

1. Was the system used?
2. For what was the system used?
3. Were the users satisfied?
4. How well did they use the system?
5. What factors were associated with successful or unsuccessful use of the system?
6. Did the system have an impact on the user's task?

A simpler means for organizing the results of evaluation, however, groups approaches and studies into those which are system-oriented, i.e., the focus of the evaluation is on the IR system, and those which are user-oriented, i.e., the focus is on the user.

5.1 System-Oriented

There are many ways to evaluate the performance of IR systems, the most widely used of which are the relevance-based measures of recall and precision. These measures quantify the number of relevant documents retrieved by the user from the database and in his or her search. They make use of the number of relevant documents (Rel), retrieved documents (Ret), and retrieved documents that are also relevant (Retrel). *Recall* is the proportion of relevant documents retrieved from the database:

$$\text{Recall} = \frac{\text{Retrel}}{\text{Rel}}. \tag{5}$$

In other words, recall answers the question, For a given search, what fraction of all the relevant documents have been obtained from the database?

One problem with (5) is that the denominator implies that the total number of relevant documents for a query is known. For all but the smallest of databases, however, it is unlikely, perhaps even impossible, for one to succeed in identifying all relevant documents in a database. Thus most studies use the measure of *relative recall*, where the denominator is redefined to represent the number of relevant documents identified by multiple searches on the query topic.

Precision is the proportion of relevant documents retrieved in the search:

$$\text{Precision} = \frac{\text{Retrel}}{\text{Ret}}. \tag{6}$$

This measure answers the question, For a search, what fraction of the retrieved documents are relevant?

One problem that arises when one is comparing systems that use ranking versus those that do not is that nonranking systems, typically using Boolean searching, tend to retrieve a fixed set of documents and as a result have fixed points of recall and precision. Systems with relevance ranking, on the other hand, have different values of recall and precision depending on the size of the retrieval set the system (or the user) has chosen to show. For this reason, many evaluators of systems featuring relevance ranking will create a recall-precision table (or graph) that identifies precision at various levels of recall. The "standard" approach to this was defined by Salton [38], who pioneered both relevance ranking and this method of evaluating such systems.

System-oriented evaluation is usually carried out with test collections, which contain a fixed collection of documents, topics, and relevance judgments for which documents are relevant to each topic. All of the above metrics are then averaged for the topics in the test collection to determine the performance of a system. One goal with test collections is to find a single measure that can characterize such performance. The emerging candidate for this metric has been *mean average precision* (MAP) [8]. To calculate MAP, average precision is calculated for each topic, and the mean of these average precision values is MAP. Average precision for a topic is calculated by averaging precision at every point where a relevant document is obtained, with values of 0 added for relevant documents not retrieved at all. This gives an average precision for each topic and MAP is calculated by averaging these points for the whole collection of topics.

No discussion of system-oriented IR evaluation can ignore the *Text REtrieval Conference* (TREC, trec.nist.gov) organized by the U.S. National Institute for Standards and Technology (NIST, www.nist.gov) [50]. Started in 1992, TREC has provided a testbed for evaluation and a forum for presentation of results. TREC is organized as an annual event at which the tasks are specified and queries and documents are provided to participants. Participating groups submit "runs" of their systems to NIST, which calculates the appropriate performance measure, usually recall and precision. TREC is organized into tracks geared to specific interests. Voorhees has grouped the tracks into general IR tasks:

- Static text – Ad Hoc
- Streamed text – Routing, Filtering
- Human in the loop – Interactive
- Beyond English (cross-lingual) – Spanish, Chinese, and others
- Beyond text – OCR, Speech, Video
- Web searching – Very Large Corpus, Web
- Answers, not documents – Question-Answering
- Retrieval in a domain – Genomics

Relevance-based measures have their limitations. While no one denies that users want systems to retrieve relevant articles, it is not clear that the quantity of relevant documents retrieved is the complete measure of how well a system performs [22, 46]. Hersh [23] has noted that medical users are unlikely to be concerned about these measures when they simply seek an answer to a clinical question and are able to do so no matter how many other relevant documents they miss (lowering recall) or how many nonrelevant ones they retrieve (lowering precision).

What alternatives to relevance-based measures can be used for determining performance of individual searches? Many advocate that the focus of evaluation put more emphasis on user-oriented studies, particularly those that focus on how well users perform real-world tasks with IR systems. Some of these studies are described in the next section.

5.2 User-Oriented

As noted above, system-oriented evaluation is valuable for comparing IR systems and algorithms, but not provide insight into how effectively they are used by their intended users. An exhaustive treatise of user-oriented research is beyond the scope of this chapter, but we can highlight the thread of user-oriented evaluation that has looked at how well users complete the types of tasks for which IR systems are intended.

In some early work, Egan et al. [15] evaluated the effectiveness of the Superbook application by assessing how well users could find and apply specific information. Mynatt et al. [36] used a similar approach in comparing paper and electronic versions of an online encyclopedia, while Wildemuth et al. [53] assessed the ability of students to answer testlike questions using a medical curricular database. For several years, TREC featured an Interactive Track that used a task-oriented approach. Results from this track showed that some algorithms found effective using system-oriented, relevance-based evaluation measures did not maintain that effectiveness in experiments with real users [24]. Another thread of user-oriented research looked as factors associated with successful use of medical IR systems [26, 27]. Similar to the TREC Interactive Track, these studies found that user success at tasks was not necessarily correlated with recall, precision, or related measures.

6 IR Techniques for LBD

Although several other chapters in this volume explore techniques for LBD, this section will present an overview of the major methods for using IR techniques in LBD. Most work in this area attempts to replicate LBD discoveries, in particular Swanson's connections of fish oil and Raynaud's [45], migraine and magnesium [47], and Somatomedin C and arginine [48]. If the LBD process can be replicated using automated methods, then the generation of new possible connections can generate hypotheses for biological researchers.

The simplest use of IR techniques involve co-occurrence of words and phrases in the text [20, 33]. Additional work has explored augmenting the approach with latent semantic indexing [19]. This technique has also been shown to be effective in drawing together ideas on the Web [21]. Other researchers have explored clustering techniques to augment this approach [44].

Other approaches have extended this statistical approach to use MeSH [43] and UMLS Metathesarus terms [51]. Hristovski and colleagues have also used UMLS Metathesarus terms, combining them with association rules to facilitate discovery [30]. Their technique has recently been extended to identify genes that play possible roles in disease [29]. Other authors have also begun to explore gene-disease connections as well [54].

These techniques have made a modest number of discoveries since replicating Swanson's original findings, indicating that new approaches will be needed to augment additional discovery. While new techniques may be developed, progress is more likely to come from the expanding biomedical knowledge, particularly in areas like functional genomics and pharmacogenomics. Although new algorithms may emerge, the growing explosion of biotechnology is more likely to provide substrate for existing methods.

7 Conclusions

IR systems have become ubiquitous for computer users from all walks of life. No productive scientist in the twenty-first century can avoid their use, and they are particularly crucial for LBD. While IR systems are widespread and easy to access, there are still challenges to their optimal use from both the technical (e.g., indexing and retrieval) and societal (e.g., open access publishing) spheres.

Acknowledgements The author's research has been generously funded by the National Library of Medicine, Agency for Healthcare Quality and Research, and National Science Foundation over the years. He is particularly grateful to the NLM for its strong leadership in promoting research and education in the field of biomedical informatics.

Suggested Readings

Baeza-Yates, R. and Ribeiro-Neto, B., eds. (1999). *Modern Information Retrieval.* New York: McGraw-Hill
A book surveying most of the automated approaches to information retrieval

Frakes, W.B. and Baeza-Yates, R. (1992). *Information Retrieval: Data Structures & Algorithms.* Englewood Cliffs, NJ: Prentice-Hall
A textbook on implementation of information retrieval systems. Covers all of the major data structures and algorithms, including inverted files, ranking algorithms, stop word lists, and stemming. There are plentiful examples of code in the C programming language

Hersh, W.R. (2003). *Information Retrieval, A Health and Biomedical Perspective* (Second Edition). New York: Springer, Berlin Heidelberg New York
A textbook on information retrieval systems in the health and biomedical domain that covers the state of the art as well as research systems

Salton, G. (1991). Developments in automatic text retrieval, *Science*, 253: 974–980
The last succinct exposition of word-statistical retrieval systems from the person who originated the approach

Voorhees, E. and Harman, D., eds. (2005). *TREC: Experiment and Evaluation in Information Retrieval.* Cambridge, MA: MIT Press
Overview of the TREC initiative

References

1. Anonymous (2007). PubMed Help, National Library of Medicine. http://www.ncbi.nlm.nih.gov/books/bv.fcgi?rid=helppubmed.chapter.pubmedhelp
2. Aronson, A., Bodenreider, O., et al. (2000). The NLM indexing initiative. *Proceedings of the AMIA 2000 Annual Symposium*, Los Angeles, CA: Hanley & Belfus, pp. 17–21
3. Bahls, C., Weitzman, J., et al. (2003). Biology's models. *The Scientist.* June 2, 2003. 5. http://www.the-scientist.com/yr2003/jun/feature_030602.html
4. Beckett, D., Miller, E., et al. (2000). Using Dublin Core in XML. Dublin Core Metadata Initiative. http://dublincore.org/documents/dcmes-xml/. Accessed 1 July 2002
5. Berners-Lee, T., Lassila, O., et al. (2001). The semantic web. *Scientific American*, 284(5): 34–43. http://www.scientificamerican.com/article.cfm?articleID=00048144-10D2-1C70-84A9809EC588EF21& catID=2
6. Borgman, C. (1999). What are digital libraries? Competing visions. *Information Processing and Management*, 35: 227–244
7. Brin, S. and Page, L. (1998). The anatomy of a large-scale hypertextual web search engine. *Computer Networks*, 30: 107–117
8. Buckley, C. and Voorhees, E. (2005). *Retrieval System Evaluation*, in Voorhees, E. and Harman, D., eds. *TREC: Experiment and Evaluation in Information Retrieval.* Cambridge, MA: MIT Press, pp. 53–75
9. Bult, C., Blake, J., et al. (2004). The Mouse Genome Database (MGD): integrating biology with the genome. *Nucleic Acids Research*, 32: D476–D481

10. Caruso, D. (2000). Digital commerce: if the AOL-Time Warner deal is about proprietary content, where does that leave a noncommercial directory it will own? *New York Times*. January 17, 2000
11. Charen, T. (1976). *MEDLARS Indexing Manual, Part I: Bibliographic Principles and Descriptive Indexing, 1977*. Springfield, VA: National Technical Information Service
12. Charen, T. (1983). *MEDLARS Indexing Manual, Part II*. Springfield, VA: National Technical Information Service
13. Coletti, M. and Bleich, H. (2001). Medical subject headings used to search the biomedical literature. *Journal of the American Medical Informatics Association*, 8: 317–323
14. Diehn, M., Sherlock, G., et al. (2003). SOURCE: a unified genomic resource of functional annotations, ontologies, and gene expression data. *Nucleic Acids Research*, 31: 219–223
15. Egan, D., Remde, J., et al. (1989). Formative design-evaluation of superbook. *ACM Transactions on Information Systems*, 7: 30–57
16. Fox, C. (1992). *Lexical Analysis and Stop Lists*, in Frakes, W. and Baeza-Yates, R., eds. *Information Retrieval: Data Structures and Algorithms*. Englewood Cliffs, NJ: Prentice-Hall, pp. 102–130
17. Frakes, W. (1992). *Stemming Algorithms*, in Frankes, W. and Baeza-Yates, R., eds. *Information Retrieval: Data Structures and Algorithms*. Englewood Cliffs, NJ: Prentice-Hall, pp. 131–160
18. Funk, M. and Reid, C. (1983). Indexing consistency in MEDLINE. *Bulletin of the Medical Library Association*, 71: 176–183
19. Gordon, M. and Dumais, S. (1998). Using latent semantic indexing for literature-based discovery. *Journal of the American Society for Information Science and Technology*, 49: 674–685
20. Gordon, M. and Lindsay, R. (1996). Toward discovery support systems: a replication, re-examination, and extension of Swanson's work on literature-based discovery of a connection between Raynaud's and fish oil. *Journal of the American Society for Information Science and Technology*, 47: 116–128
21. Gordon, M., Lindsay, R., et al. (2002). Literature-based discovery on the World Wide Web. *ACM Transactions on Internet Technology*, 2: 261–275
22. Harter, S. (1992). Psychological relevance and information science. *Journal of the American Society for Information Science*, 43: 602–615
23. Hersh, W. (1994). Relevance and retrieval evaluation: perspectives from medicine. *Journal of the American Society for Information Science*, 45: 201–206
24. Hersh, W. (2001). Interactivity at the Text Retrieval Conference (TREC). *Information Processing and Management*, 37: 365–366
25. Hersh, W. (2003). *Information Retrieval: A Health and Biomedical Perspective* (Second Edition). Berlin Heidelberg New York: Springer. http://www.irbook.info
26. Hersh, W., Crabtree, M., et al. (2002). Factors associated with success for searching MEDLINE and applying evidence to answer clinical questions. *Journal of the American Medical Informatics Association*, 9: 283–293
27. Hersh, W., Crabtree, M., et al. (2000). Factors associated with successful answering of clinical questions using an information retrieval system. *Bulletin of the Medical Library Association*, 88: 323–331
28. Hersh, W. and Hickam, D. (1998). How well do physicians use electronic information retrieval systems? A framework for investigation and review of the literature. *Journal of the American Medical Association*, 280: 1347–1352
29. Hristovski, D., Peterlin, B., et al. (2005). Using literature-based discovery to identify disease candidate genes. *International Journal of Medical Informatics*, 74: 289–298
30. Hristovski, D., Stare, J., et al. (2001). Supporting discovery in medicine by association rule mining in Medline and UMLS. *MEDINFO 2001 – Proceedings of the Tenth World Congress on Medical Informatics*, London, UK: IOS Press, pp. 1344–1348
31. Langville, A. and Meyer, C. (2006). *Google's PageRank and Beyond: The Science of Search Engine Rankings*. Princeton, NJ: Princeton University Press
32. Lawrence, S., Giles, C., et al. (1999). Digital libraries and autonomous citation indexing. *Computer*, 32: 67–71

33. Lindsay, R. and Gordon, M. (1999). Literature-based discovery by lexical statistics. *Journal of the American Society for Information Science and Technology*, 50: 574–587
34. Miller, E. (1998). An introduction to the resource description framework. *D-Lib Magazine*, 4. http://www.dlib.org/dlib/may98/miller/05miller.html
35. Miller, N., Lacroix, E., et al. (2000). MEDLINEplus: building and maintaining the National Library of Medicine's consumer health web service. *Bulletin of the Medical Library Association*, 88: 11–17
36. Mynatt, B., Leventhal, L., et al. (1992). Hypertext or book: which is better for answering questions? *Proceedings of Computer-Human Interface 92*. 19–25
37. Perkel, J. (2003). Feeding the info junkies. *The Scientist*. June 2, 2003. 39. http://www.the-scientist.com/yr2003/jun/feature14_030602.html
38. Salton, G. (1983). *Introduction to Modern Information Retrieval*. New York: McGraw-Hill
39. Salton, G. (1991). Developments in automatic text retrieval. *Science*, 253: 974–980
40. Salton, G., Fox, E., et al. (1983). Extended Boolean information retrieval. *Communications of the ACM*, 26: 1022–1036
41. Salton, G. and Lesk, M. (1965). The SMART automatic document retrieval system: an illustration. *Communications of the ACM*, 8: 391–398
42. Spitzer, V., Ackerman, M., et al. (1996). The visible human male: a technical report. *Journal of the American Medical Informatics Association*, 3: 118–130
43. Srinivasan, P. (2004). Text mining: generating hypotheses from MEDLINE. *Journal of the American Society for Information Science and Technology*, 55: 396–413
44. Stegmann, J. and Grohmann, G. (2003). Hypothesis generation guided by co-word clustering. *Scientometrics*, 56: 111–135
45. Swanson, D. (1986). Fish oil, Raynaud's syndrome, and undiscovered public knowledge. *Perspectives in Biology and Medicine*, 30: 7–18
46. Swanson, D. (1988a). Historical note: information retrieval and the future of an illusion. *Journal of the American Society for Information Science*, 39: 92–98
47. Swanson, D. (1988b). Migraine and magnesium: eleven neglected connections. *Perspectives in Biology and Medicine*, 31: 526–557
48. Swanson, D. (1990). Somatomedin C and arginine: implicit connections between mutually isolated literatures. *Perspectives in Biology and Medicine*, 33: 157–186
49. van Rijsbergen, C. (1979). *Information Retrieval*. London: Butterworth
50. Voorhees, E. and Harman, D., eds. (2005). *TREC: Experiment and Evaluation in Information Retrieval*. Cambridge, MA: MIT Press
51. Weeber, M., Vos, R., et al. (2003). Generating hypotheses by discovering implicit associations in the literature: a case report of a search for new potential therapeutic uses for thalidomide. *Journal of the American Medical Informatics Association*, 10: 252–259
52. Weibel, S. (1996). The Dublin Core: a simple content description model for electronic resources. *ASIS Bulletin*, 24(1): 9–11. http://www.asis.org/Bulletin/Oct-97/weibel.htm
53. Wildemuth, B., deBliek, R., et al. (1995). Medical students' personal knowledge, searching proficiency, and database use in problem solving. *Journal of the American Society for Information Science*, 46: 590–607
54. Wren, J., Bekeredjian, R., et al. (2004). Knowledge discovery by automated identification and ranking of implicit relationships. *Bioinformatics*, 20: 389–398

Biomedical Application of Knowledge Discovery

A. Koike

Abstract With rapid progress in biomedical fields, the knowledge accumulated in scientific papers has increased significantly. Most of these papers draw only a fragmental conclusion from the viewpoint of scientific facts, so discovery of hidden knowledge or hypothesis generation by leveraging this fragmental information has come into the limelight and more expectations on the system constructions to assist them has been paid. To respond to these expectations, we have developed a system called BioTermNet (http://btn.ontology.ims.u-tokyo.ac.jp:8081/) to make a conceptual network by connecting conceptual relationships (fragmental information) explicitly described in papers and explore the hidden relationships in the conceptual network. The conceptual relationships are extracted by hybrid methods of information extraction and information-retrieval techniques. This system has a potential for wide application. After the validation of system performance, we take up some topics of conceptual network-based analysis and refer to other applications in the future prospects section.

1 Overview of BioTermNet

After Swanson's prediction of the dietary effect of fish oil as a pioneering work of knowledge discovery in the biomedical field [1], several knowledge discovery methods/systems were proposed [2–5], and some of them were introduced in other sections of this book. Most knowledge discovery systems use the simple ABC model

A. Koike
Central Research Laboratory, Hitachi Ltd. 1-280 Higashi-Koigakubo Kokubunji City, Tokyo, 185-8601, Japan
and
Department of Computational Biology, Graduate School of Frontier Science, The University of Tokyo, Kiban-3A1(CB01) 5-1-5, Kashiwanoha Kashiwa, Chiba 277-8561, Japan
akoike@cb.k.u-tokyo.ac.jp

P. Bruza and M. Weeber (eds.), *Literature-based Discovery*,
Springer Series in Information Science and Knowledge Management 15.

proposed by Swanson, which assumes that if a paper describes the relationship between concepts/terms A and B, and another paper describes the relationship between concepts B and C, the relationship between concepts A and C is inferred, even if there are no papers that explicitly describe the relationship between A and C. Although our method is based on this idea, it has the following characteristics to develop a system that effectively predicts hidden relationships and proposes evidence for them so that users can easily comprehend the prediction results.

(1) Originally developed dictionaries with publicly available thesauruses to consider multiwords, synonyms, hyponyms, and hypernyms are utilized.
(2) The uses of a statistical approach and syntactic analysis for extraction of conceptual relationships are combined.
(3) Semantic types such as genes and diseases are added to enable intermediate candidate concepts to be focused on.
(4) Multiple intermediate steps (AB_1B_2C model and $AB_1B_2B_3C$ model, etc.) can be used.
(5) Evidential sentences for syntactic analysis and evidential abstracts for the statistical approach are presented.
(6) Multiple start/end points can be used.
(7) These discovery processes are fully automatically performed. (without manual selections of intermediate concepts).

Characteristic (1) is down-to-earth but has a large effect on knowledge discovery performance. There are some methods for extrapolating terms by statistical methods in order to extract the relationships among synonyms, hypernyms, and hyponyms. The trivial error of recognition of concepts/terms leads to the extraction of meaningless hidden relationships, so manual checks of automatically extracted terms are necessary in most cases. Also in our study, a manual check is performed after the automatic term extraction before the registration of these terms into our dictionaries. In BioTermNet, a public thesaurus: Unified Medical Language System (UMLS), disease ontology, OMIM, specific thesauruses developed in our laboratory: GENA (gene name dictionary), family name dictionary, and other semi-automatically extracted terms to compensate for inadequacy of semantic classes of UMLS such as pathway name class, are used.

The syntactic-analysis and statistical approaches in (2) compensate for the deficiencies of each other. If the statistical approach is used, important but minor relationships, especially newly discovered relationships, tend to be ignored. Furthermore, distinguishing edge kinds such as protein interactions and functionally related proteins in taking account of a signaling/metabolic network such as "protein-A activates protein-B and protein-B inhibits protein-C" is sometimes necessary. Edge kinds are only distinguished by syntactic analysis. If only syntactic analysis is applied, on the other hand, extracting meaningful two-concept relationships for all semantic types with high performance is quite difficult. This is because the expressions of two-concept relatedness are quite various, some relationships are described over more than one sentence, and information extraction for complicated sentences specific in the biomedical domain is quite an abstruse task for current

NLP techniques. In this system, interaction/regulation relationships between a protein/gene/family/compound (drug) or a gene/protein-biological process function are extracted by syntactic analysis. That is because they are constitutive to consider a pathway network, and their relationships can be more clearly described than other two-concept-type relationships.

Semantic restriction in (3) is effective for considering a user's idea such that the final concept is induced from start concept by mediating chemical compounds in one step or by mediating genes and chemical compounds in two steps. Multiple intermediate steps in (4) reduce the gap of conceptual relatedness and lead to the knowledge discovery of easily understandable relationships. The presentation of evidence in (5) is a fundamental function because the researcher cannot determine whether the intermediate concept is a meaningful candidate without evidential documents or adequate background knowledge. Various contrivances of other systems are summarized in [6] below and in other sections of this book. By the extension of multiple start/end concepts in (6), the application of our system to analysis of high-throughput results becomes possible.

The objective of BioTermNet is not just only knowledge discovery but to present hidden relationships and information to enable users to confirm or investigate the relationship actually described in the paper. Both the "open discovery process" (only start concept is given) and the "closed discovery process" (start and end concepts are given) can be performed in our system.

2 System and Methods

BioTermNet has been constructed using the following steps to explore hidden relationships:

- Step 1: Term Recognition
- Step 2: Calculation of conceptual relationships based on syntactic analysis: gene/protein/family/compound interaction and gene-function extraction
- Step 3: Calculation of conceptual relationships based on statistical analysis
- Step 4: Calculation of conceptual network based on syntactic and/or statistical relationships
- Step 5: Drawing of conceptual network

Steps 1–3 are pre-calculated. After that, the queries are given by users and steps 4 and 5 are interactively processed. When query q (in (2) described below) consists of multiple concepts or some conditions are imposed in the calculation of relationship (for example, the search of related genes of a certain gene under the cell division), all concept IDs and their frequencies per abstract are indexed to interactively calculate the conceptual relationships instead of the pre-calculation of step 3.

In step 1, all concepts in MEDLINE abstracts are converted into concept IDs: recognition of terminologies, resolution of ambiguities (for example, GCK is an abbreviation of both glucokinase and germinal center kinase), and addition of semantic classes (basically UMLS semantic classes) for each terminology are performed. There are trivial variations of terminologies such as "NF-kappa B, NF kappaB" and "Alzheimer disease, Alzheimer's disease" are absorbed in these recognition steps as far as possible using a devised trie, as described in a previous work [7].

In step 2, relationships explicitly described in abstracts are extracted by syntactic analysis. After the recognition of terminology, the sentence is shallowly parsed, noun phase bracketing is done, sentence structure is analyzed, and then, the ACTOR and OBJECT relationship is recognized. The extractions of ACTOR-OBJECT relationships are then performed when they are described as being in a certain relationship (such as *inhibition* of OBJECT by ACTOR, ACTOR *regulates* OBJECT). In this system, PRIME data [8–10] were used for the syntactic-analysis approach. They were extracted from MEDLINE abstracts for each species, such as "*Saccharomyces cerevisiae*," "*Caenorhabditis elegans*," "*Drosophila melanogaster*," "mice," "rats," and "humans." These included 920,000 (nonredundant) protein interactions and 360,000 annotated gene-function relationships for major eukaryotes as of April 2006. Details on information extraction have been described in previous papers [8–10].

In step 3, there are several statistical methods to extract conceptual relationships and their performance is compared in the section "3. Comparison of Calculation Methods." In BiotermNet, related concepts were calculated by applying a similar text-search technique, the vector space model [11], which is widely recognized to effectively select representative concepts. Equation (1) represents the weighting for document d with query q and is called Lnu weighting. Lnu weighting is similar to TF-IDF but is recognized to be superior in considering the document length and in rendering the "verbosity" of a document. The importance of taking into account the effect of document length on MEDLINE abstracts has been presented in previous work [12]. Equations (2) and (3) represent the weights for query term t_i in the queries and that in document d, respectively. Equation (4) calculates the "representativeness" of concept p in documents including query q, and the statistical relatedness of query q to concept p is derived. When documents including query term q; D(q), exceed 30,000 in our system, the top 30,000 weighted documents defined by (1) are selected in interactive mode in step 3 and are used as D(q) to reduce the CPU calculation time and interactively present the conceptual network.

$$w(d,q) = \frac{\sum_i wq(t_i|q) * wd(t_i|d)}{L + \kappa * [dlen(d) - L]}, \tag{1}$$

$$wq(t_i,q) = \frac{1 + \log[tf(t_i|q)]}{1 + \log[tf(.|q)]}, \tag{2}$$

$$wd(t_i,d) = \frac{1 + \log[tf(t_i|d)]}{1 + \log[tf(.|d)]} * \{1 + \log(N/df(t_i))\}, \text{ and} \tag{3}$$

$$rel(p,q) = \left\{ \sum_{d \in D(q)} \frac{1 + \log[tf(p|d)]}{1 + \log[tf(.|d)]} \right\} * \{1 + \log(N/df(p)\}. \tag{4}$$

Here, *dlen(d)* is the number of different concepts in document d, L is the average *dlen(d)* over all documents, and κ in (1) is a slope constant set to 0.2. $df(t_i)$ is the number of documents including t_i, and N is the total number of documents. $tf(t_i|q)$ is the frequency of concept t_i in query q (consisting of multiple concepts), and $tf(.|q)$ is their average frequency over all concepts in query q. In a default, a query consists of a unique concept ID, so (2) equals 1. $tf(t_i|d)$ is the frequency of t_i appearing in document d, and $tf(.|d)$ is the average frequency over all concepts in document d.

When all steps are calculated using the statistical approach, the score in each step (score(p,q): score between concepts p and q) is defined by log(*rel(p,q)/rel(q,q)*). The value of (4) depends on the number of retrieved documents, so normalization is performed by the relatedness query itself *rel(q,q)*. When only syntactic analysis or syntactic analysis and statistical approaches are combined, each step based on syntactic analysis is set to log(0.95) as a default to prioritize syntactic analysis data. The priority of the statistical relationship and syntactic analysis relationship depends on the research purpose, so that priority can be changed to the average relatedness for gene/protein/family/compound interaction and that for gene-function with options. When multiple intermediate steps are used, the sum or harmonic mean of logarithms of the score of each step is defined as the path score in both closed and open discovery systems. The calculated path score is assigned to each node. When there are multiple paths for a node, a maximum score is assigned to the node. The generated network changes depending on the path score calculation method.

2.1 Closed Discovery System

Start Query (Qs) $\rightarrow$ B and end query (Qe) $\rightarrow$ B retrieval (two-step network), Qs $\rightarrow B_1$, $B_1 \rightarrow B_2$, Qe $\rightarrow B_2$ (three-step network), Qs $\rightarrow B_1$, $B_1 \rightarrow B_2$, Qe $\rightarrow B_3$, $B_3 \rightarrow B_2$ (four-step network), are available steps in a closed discovery process with a single start/end query. Although adding intermediate steps is also theoretically possible, the web service is limited to a four-step network. That is because more than four-step discovery requires too much CPU-calculation time to ensure efficient discovery of knowledge interactively. Knowledge discovery results based on too many intermediate steps tend to generate spurious relationships. The semantic type in each step (layer) can be specified.

Representative/related concepts that are greater than a certain threshold extracted by start query and end query are pooled as pseudo-intermediate concepts in the two-step network. Common representative concepts are selected from these and the path score (the sum or harmonic mean of the logarithm of each relationship log(score(p,q)) as mentioned above) passing over each pseudo-intermediate concept is calculated. Intermediate concepts having the top 10–50 (user choice) path scores are selected and their network is presented.

In the three-step network, pseudo-intermediate concepts B_1 are selected and used as queries to explore the next pseudo-intermediate representative concepts B_2. B_2

concepts connecting Qe concepts with the highest 10–50 path scores are selected and the B_1 concepts, which connect to the selected B_2 concepts and whose highest path scores are more than the lowest path scores of selected B_2 concepts, are selected.

In the four-step network, pseudo-intermediate concepts (B_1 and B_3) are used as queries. Similarly, B_2 concepts connecting B_1 and B_3 concepts with the top 10–50 (user choice) path scores are selected, and B_1 and B_3 concepts, which are connected to the selected B_2 concepts and whose highest path scores are more than the lowest path scores of the selected B_2 concepts, are selected.

When closed discovery system is extended to multiple start and end queries or multiple start queries with a single end query, intermediate concepts are selected by voting from the connecting start and/or end query concepts instead of the max path score. For example, start queries: Qs^1, $Qs^2, \ldots,$ Qs^N, and end queries: Qe^1, $Qe^2, \ldots,$ Qe^M are given for three-step discovery, after the calculation of $Qs^n \rightarrow B_1{}^n$, $B_1{}^n \rightarrow B_2{}^n$, and $Qe^n \rightarrow B_2{}^n$ for each start and end query. Then, B_2 concepts are selected in descending order of the sum of the number of connecting start concepts and end concepts (if there are a unique end concept, the number of connecting start concepts is used) under the condition that B_2 connects B_1 and Qe. Then, the B_1 concepts, which connect to the selected B_2 concepts and whose highest path scores are more than the lowest path scores of selected B_2 concepts, are selected.

2.2 Open Discovery System

$Qs \rightarrow B$, $B \rightarrow C$ (two-step network), $Qs \rightarrow B_1$, $B_1 \rightarrow B_2$, $B_2 \rightarrow C$ (three-step network), and $Qs \rightarrow B_1$, $B_1 \rightarrow B_2$, $B_2 \rightarrow B_3$, $B_3 \rightarrow C$ (four-step network) are provided in this system as an open discovery process with a single start concept. The semantic types in each step can be specified. All path scores are calculated, which is similar to the closed discovery system. Three methods are applied to select C concepts. The first is (1) Max score method: the descending order of path scores for C concepts, and the second is (2) Voting method: the voting of C concepts in the final step ($B \rightarrow C$), i.e., how many times the target concept is regarded as a concept related to intermediate terms. (For example, when $B^1 \rightarrow C^1$, $B^2 \rightarrow C^1$, $B^3 \rightarrow C^2$, and $B^4 \rightarrow C^1$ are calculated, C^1 is counted as 3 and C^2 is counted as 1.) The third is (3) Sum method: the descending order of the sum of the path score for C concepts. As a default, the max score method, which is superior to other methods in most cases in our preliminary study using Lnu weighting, is available on the Web.

When multiple start concepts are given, after the calculation of each discovery process, the final concept is selected in descending order of the number of connecting start concepts that are used. After that, intermediate concepts, which directly/indirectly connect to the selected final concepts and whose highest path scores are more than the lowest path scores of selected final concepts, are selected.

Although the details are not mentioned here, the number of pseudo-intermediate concepts is reduced in each concept expansion step using the fragmental maximal path score to efficiently draw a network in both open and closed discovery on the Web service [6].

3 Comparison of Calculation Methods

As described above, BioTermNet uses both syntactic analysis and statistical method for extraction of explicit relationships. In the latter, concepts that noticeably co-occur in abstracts (or a document unit) are assumed to be related, but it is not assured that an explicit relationship between them is described in the text. Therefore, a statistical method that scores explicitly described relationships higher than mere co-occurrence terms (relationships) is desirable. The required specifications for the calculation method are as follows: (1) it must score explicit relationships highly; and (2) it must give an appropriate score to calculate the network and assist in plausible new discovery. If only the performance of knowledge discovery is pursued, only (2) is required, but (1) is set to the essential qualification to present easily understandable evidence. To select the best calculation method from these two points, we evaluated eight statistical methods and SVD (Singular Value Decomposition).

The Dice coefficient (5), Cosine coefficient (6), Simpson coefficient (7), and Mutual Information (MI) (8) are generally applied to identify the relatedness between two terms/concepts. MI tends to evaluate low-frequency concepts too highly, so the frequency of documents containing both p and q, is used as a multiplication factor in MI* freq (9). TF-IDF (10) is a widely used term weight in text retrieval. Although HyperGsum (HG) based on hypergeometric distribution is not widely known, it is superior at representing characteristic terms in the retrieved documents [13]. Each method is defined as follows.

Dice coefficient

$$rel(p,q) = \frac{2*df(p,q)}{df(p)+df(q)}. \tag{5}$$

Cosine coefficient

$$rel(p,q) = \frac{df(p,q)}{\sqrt{|df(p)||df(q)|}}. \tag{6}$$

Simpson coefficient

$$rel(p,q) = \frac{df(p,q)}{\min(df(p),df(q))}. \tag{7}$$

Mutual information

$$rel(p,q) = \log\left(\frac{N*df(p,q)}{df(p)*df(q)}\right). \tag{8}$$

MI*freq

$$rel(p,q) = df(p,q) * \log\left(\frac{N * df(p,q)}{df(p) * df(q)}\right). \tag{9}$$

TF-IDF

$$rel(p,q) = tf(p|D(q)) * \log(N/df(p)). \tag{10}$$

HyperGsum (HG)

$$rel(p,q) = -\log(hgs(N, df(p), df(q), df(p|D(q))),$$

$$hgs(N, df(p), df(q), df(p|D(q))) = \sum_{l \geq df(p|D(q))} \frac{C(df(p), df(p|D(q))C(N - df(p), df(q) - df(p|D(q))}{C(N, df(q))}. \tag{11}$$

Here, *df(p)* is the frequency of documents including concept p; *df(p,q)* is the frequency of documents including concepts p and q; *D(q)* is the set of documents containing concept q; $tf(p|D(q))$ is the frequency of concept p in documents *D(q)*; *N* is the number of all documents; and $C(t,u) =_t C_u = u!/t!(t-u)!$

Latent semantic indexing (LSI) based on singular vector decomposition (SVD) reduces the number of dimensions of the term-document space, and related concepts or similar documents are projected in similar positions in the reduced space [14]. SVD is widely used in text retrieval systems to automatically consider synonyms of query terms. Rectangular term (concept ID)-document matrix **D** (matrix element D_{ij} is the frequency of the i-th concept ID in the j-th document, D is a term-by-n document matrix) is decomposed as

$$\mathbf{D} = \mathbf{U}\Sigma\mathbf{V}^T, \tag{12}$$

where U is a m-by-m unitary matrix and its columns consist of eigenvectors of $\mathbf{D} \cdot \mathbf{D}^T$ (called left singular vectors) and V is a n-by-n unitary matrix and its columns consist of eigenvectors of $\mathbf{D}^T \cdot \mathbf{D}$ (called right singular vectors). Σ is a diagonal matrix of non-negative singular values, and those values are in descending order. This is known as the singular value decomposition of matrix **D**.

Concept vector **q** is approximated by k dimensions using the first k columns of the **U** matrix and the highest k singular values of Σ:

$$\hat{\mathbf{q}}^{(k)} = \mathbf{q}^T \mathbf{U}_k \Sigma_k^{-1}. \tag{13}$$

The relatedness between two concepts, p and q, is represented by the cosine of their vectors in the k-approximation of matrix **D**:

$$rel(p,q) = \frac{\hat{\mathbf{p}}^{(k)T} \cdot \hat{\mathbf{q}}^{(k)}}{||\hat{\mathbf{p}}^{(k)T}|| \, ||\hat{\mathbf{q}}^{(k)}||}. \tag{14}$$

The use of log entropy weighting represented by (15) instead of concept frequency in matrix D has been reported to be effective [15]. By the introduction of log entropy, low frequency concepts are emphasized and high-frequency concepts are lowly weighted.

$$\begin{aligned} D_{ij} &= l_{ij} * g_{ij}, \\ &= \log_2(1+f_{ij}) * \left\{1 + \left(\frac{\sum_j p_{ij} \log_2(p_{ij})}{\log_2 n}\right)\right\}, \\ p_{ij} &= \frac{f_{ij}}{\sum_j f_{ij}}. \end{aligned} \quad (15)$$

In this study, both SVD based on the simple concept frequency matrix (hereinafter referred to as SVD) and that based on (15) (hereinafter referred to as SVD-entropy) were calculated. Calculations were performed using the SVDPACK Lanczos algorithm [16]. The dimensionality is reduced to $k = 200$ after testing various dimensions.

MEDLINE abstracts with the MeSH term "human" as of April 2006 (222,531 concepts and 8,044,562 documents) are used in the test of relationship extraction. In Fig. 1, 15,000 HPRD (human protein reference database [17]) protein–protein interactions are taken as a golden standard, and the recall of protein interactions against retrieved concepts (retrieved concepts include all semantic types) is plotted. In Fig. 1, Lnu weighting exhibits the best performance. As described in [18], Lnu weighting exhibits the best performance in gene-disease (OMIM data [19]), gene-GO (gene ontology)_biological process, and gene-GO_molecular functions [20]. Interestingly, although HyperGsum recall is low in Fig. 1, it exhibits the best recall

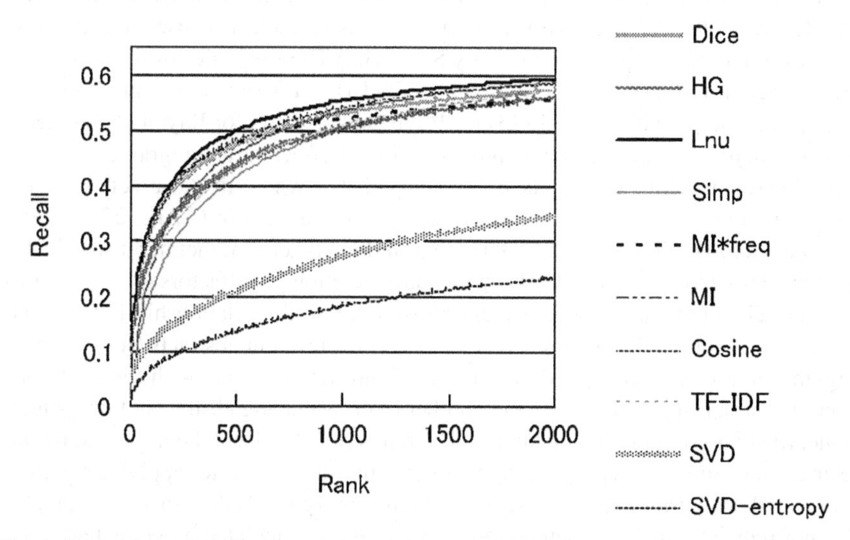

Fig. 1 Comparison of recall of protein interaction extraction at each retrieved rank among statistical methods and SVD

of the GO_cellular component [18]. All statistical methods except Lnu weighting and TF-IDF presume that each concept appears only once per abstract. The target concept frequency per abstract is close to one only in the GO_cellular component relationship, so HyperGsum seems to be better than other statistical methods when the average target concept frequency is close to one [18]. Otherwise, Lnu weighting is better than other methods from the viewpoint that desirable statistical method scores explicitly described relationships higher than mere co-occurrence concepts. SVD is well recognized to be effective in considering synonyms without dictionaries and document clustering because it reduces the document-concept space, and as a result, hidden concept relationships including synonyms can be considered. In Fig. 1, however, both SVD and SVD-entropy exhibit significantly lower performance than the other methods and it also exhibits significantly lower performance than other methods in gene-GO and gene–disease relationships. There is a method which performs the SVD based on the pseudo-documents where all abstracts describing about each gene are concatenated and are used as a document unit instead of an abstract, and estimates gene–gene distances by calculating pseudo-document distances [21]. Even if this method is applied, the performance is rather worse than the result of simple SVD in Fig. 1.

The largest difference between SVD and other methods is that SVD considers all conceptual relationships to determine the relatedness of two targeted concepts, but other statistical methods consider only the overlap of targeting a two-concept distribution in abstracts. We expect that rather than consideration of all relationships, focusing attention on a two-concept distribution deflection is important to extract explicitly described relationships in a text.

Next, to evaluate the knowledge discovery performance of each method, conventional subjects are used as golden standard, and the rank of the target concept predicted by each method is summarized in Table 1. The MEDLINE abstracts with MeSH term "human" up to the year in parenthesis in Table 1, is used. (At the next year, original predictions were made by Swanson.) The semantics of final concepts are restricted as follows: (1) the effective lipid (UMLS semantic type: T119) or biologically active substance (T123) for Raynaud's disease, for Raynaud's disease–fish oil relationship [1], (2) the compound (T109) effective for migraine, for Mg–migraine relationship [22], (3) the compounds (T109) whose intake are effective for somatomedin-C-related phenomena in vivo, for somatomedin C-Arg [23], (4) the indomethacin-applicable disease (T047), for indomethacin–Alzheimer disease relationship [24], (5) the genes (T116), which are the etiological factors of schizophrenia, for schizophrenia-Ca-independent phospholipase A2 relationship [25]. When these golden standard start and end concepts co-occurred in abstracts (for example, Mg–migraine), concepts that co-occurred (or did not co-occur) with the start concept in abstracts not greater than the number of co-occurring abstracts of the golden standard are used as prediction targets. The ranks of Table 1 will be improved by the restriction of final concept semantic types [6]. In this study, after applying the max score method, normalized max score method, voting method, sum score method, and normalized sum score method (all path scores are calculated based on harmonic mean of logarithm of each step relatedness) to select the final concept, the best and

Table 1 Summary of ranks of target concepts in open discovery process

	Dice	HG	Lnu	Simp	MI*freq	MI	Cos	TF-IDF	SVD
Raynaud's disease → fish oil (1985)	107 [192]	214 [144]	205 [175]	1,305 [206]	21 [206]	320 [149]	228 [149]	1,557 [126]	807 [174]
Migraine → Mg (1987)	131 [8]	4 [6]	40 [10]	15,540 [1]	7,273 [1]	8,732 [6]	161 [6]	19,628 [12]	21,019 [6,397]
Somatomedin C → Arg (1989)	91 [49]	198 [79]	149 [19]	9,637 [71]	370 [71]	2,026 [58]	146 [56]	716 [33]	37 [8]
Indomethacin → Alzheimer disease (1995)	383 [3]	61 [1]	256 [17]	14,929 [1]	1,950 [1]	4,879 [1]	383 [1]	8,323 [4]	16,800 [4,477]
Schizophrenia → Ca-independent phospholipase A2 (1997)	4 [71]	5 [42]	19 [46]	7,112 [42]	5,438 [42]	1,784 [37]	44 [37]	1,299 [22]	8,204 [5,529]

Values *outside* and *inside brackets* represent the results of max score method and those of voting method, respectively. Values of SVD *outside brackets* represent the ranks without intermediate concepts and those *inside brackets* represent the results based on log entropy instead of term frequencies in matrix D in (12)

second best results among all methods, the normalized max score and voting method result, are summarized in Table 1. Concerning SVD, intermediate steps have less meaning, so only the result of the rank of direct relationship is depicted.

Furthermore, an appropriate final concept is not the only provided target concept (for example, an effective lipid for Raynaud's disease is not only fish oil), so the target concept in Table 1 is not necessarily ranked first. Actually, plausible concepts are ranked higher than the correct answer concepts [6]. For example, Gitelman syndrome and SUNCT syndrome are ranked higher than Alzheimer's disease in the results of max score method with Lnu-weighting. Recently, growth failure of Gitelman syndrome was reported to be remedied by increasing the indometacin dose (PMID:10569969) and response to treatment with indomethacin in patients with SUNCT syndrome was also reported (PMID:9055807). In this sense, a simple comparison of the ranks of Table 1 is not appropriate to evaluate these methods. The performance of each method, however, can be conjectured.

As shown in Table 1, except Lnu-weighting, HyperGsum, and Dice, the voting method shows the best performance in most cases and the difference between the voting method and the normalized max score is quite large. Since the voting method tends to select the major concepts due to their characteristics, the methods with small difference between voting and max score methods, Dice, HyperGsum, and Lnu-weighting, would be desirable.

Surprisingly, HyperGsum seems to be better than Lnu-weighting in some cases, in spite of its low performance in the extraction of explicitly described relationships. This might be because the influences of the target concept density in abstracts are receded in the case of the large size of abstract sets such as those of Mg and Alzheimer's disease. The decrease in performance of Lnu-weighting in the network calculation can be expected due to the lack of symmetry, that is, the relatedness of A $\rightarrow$ B and B $\rightarrow$ A are different in this method, and because the relatedness strength of conceptual relationships is not absolute. The latter does not influence the extraction of relationships but they may be a problem for the network score calculation. Therefore, the direct comparison between relationship strengths $B_1 \rightarrow C$ and $B_2 \rightarrow C$ is not appropriate. Although the pseudo normalization was performed (score(p,q) = rel(p,q)/rel(q,q)), the normalization did not seem to be sufficient.

As shown in Table 1, the results of SVD are worse than other techniques except somatomedin C – Arg. That is, the final concepts are quite far from the start concept even in the dimensionally reduced space. Even if the intermediate concepts were introduced, the ranks were not improved (data not shown). This indicates that the desired relationships as knowledge discovery/hypothesis generation are close to the indirect relationships based on the connections of multiple facts rather than the relationships of similar concepts in latent semantic space. Although the simple term frequency based SVD shows better results than entropy based SVD in the extraction of relationships explicitly described in text as shown in Fig. 1, concerning the distance of the knowledge discovery, the results are the reverse.

Although the efficiency of a statistical method depends on the network calculation method, Lnu-weighting and HyperGsum seem to be better than other methods such as conventionally used MI, at least when the harmonic mean of the logarithm of a statistical value is applied. Lnu-weighting is used in the sections below.

The details of the BioTermNet open and closed discovery result analysis of "Mg–migraine" and "Raynaud's disease–fish oil" based on Lnu-weighting were reported in [6]. The plausible and understandable evidences were also presented [6].

4 Application to Pharmaco Genomics

Pharmaco genomics, that is, the study of SNPs, which causes personal differences in beneficial and side effects of drugs, is becoming an important subject. To specify the targeted SNPs or SNP haplotypes (the set of SNP alleles along a region of a chromosome) by SNP typing, the inquisition of the gene containing the targeting SNPs is necessary. There are various genes containing the target SNPs, for example, genes to which the compound binds, genes in the same signaling pathway of the drug targeting gene, or genes related to the metabolism of the drug. Making a list of candidate genes containing functional SNPs or haplotypes for a drug requires much background knowledge. Therefore, BioTermNet assists for searching candidate genes for SNP typing. For example, aspirin is a widely used drug with antipyretic, analgesic, anti-inflammatory, and anticoagulant effects. Low-dose long-term aspirin has an inhibitory effect on platelet aggregation and causes a prolongation of the bleeding time and is used for reducing the odds of heart disease including serious atherothrombotic vascular event. Aspirin resistance describes an inability of aspirin to produce these anticipated effects on platelet function [26]. Several possible causes of aspirin resistance have been discussed and the difference of individual genotypes is one of them. Although several experiments have been performed for searching the aspirin resistance related SNPs and haplotypes [27, 28], all aspirin resistance related SNPs and haplotypes have not yet been resolved. In Fig. 2, possible genes related to aspirin directly or indirectly from the viewpoint of bleeding time are depicted (closed discovery with first layer: gene, and second layer: compound) as a result of BioTermNet. Relationships between aspirin and first-layer genes are restricted to syntactic analysis data; so genes in the regulation relationships with aspirin are extracted. By clicking the edges, the following interpretation is obtained. "Aspirin inhibits both the PTGS1 (COX-1) and PTGS2 (COX-2) (syntactic analysis PRIME result, PMID:10096266) and PTGS1 and PTGS2 catalyze the formation of various prostaglandins and thromboxane A2 (PRIME result, PMID:12032335). The inhibition of thromboxane A2 synthesis causes the prolongation of bleeding time (PMID:2526385)." Accordingly the administration of aspirin causes the prolongation of bleeding time. As other concepts are drawn as a conceptual network, by going over the other conceptual relationships (edges), other possible interpretations are obtained. When only one intermediate layer with semantic class of gene is used for the closed discovery, the platelet glycoproteins such as GP1BA and ITGA2B are highly ranked. The interpretation of aspirin effect on bleeding time with mediating platelet glycoprotein is also obtained as follows. "Aspirin inhibits platelet glycoprotein activation (PMID:15370100) and this leads to prolong the bleeding time (PMID:16525577)." The correct interpretations of the mechanism of aspirin effect on bleeding time are not necessarily unique. Of these listed genes in closed discoveries, all genes with plausible intermediate terms and interpretations are aspirin relevant gene candidates appropriate for SNP typing.

Furthermore, when genes in the first layer are limited to the genes with functional SNPs or haplotypes, (SNPs or haplotypes, that have been reported to have any effects on disease risks and drug side effects, aspirin and polymorphism are used as a start query), it represents the priority of SNP typing of targeted genes to specify the drug-effective-related SNPs at the current knowledge although the document information itself is not sufficient. In Fig. 2, most of genes have functional SNPs or haplotypes and some of them are relevant to aspirin resistance. By incorporating SNP frequencies in an SNP database such as HapMap [29] into BioTermNet, the selection of appropriate SNP-typing target genes based on their SNP frequency data and documents knowledge is also theoretically possible.

5 Application to Linkage Analysis

In linkage analysis, the disease-causing (related) gene is explored based on the strength of linkage of markers using disease-infected patient families. In many cases, however, the located genome regions of the disease-related gene candidates are quite broad, and the predicted regions sometimes include more than 100–1,000 genes. Furthermore, the fact that the highest LOD score does not assure the indication of target disease-related gene position makes the specification of target

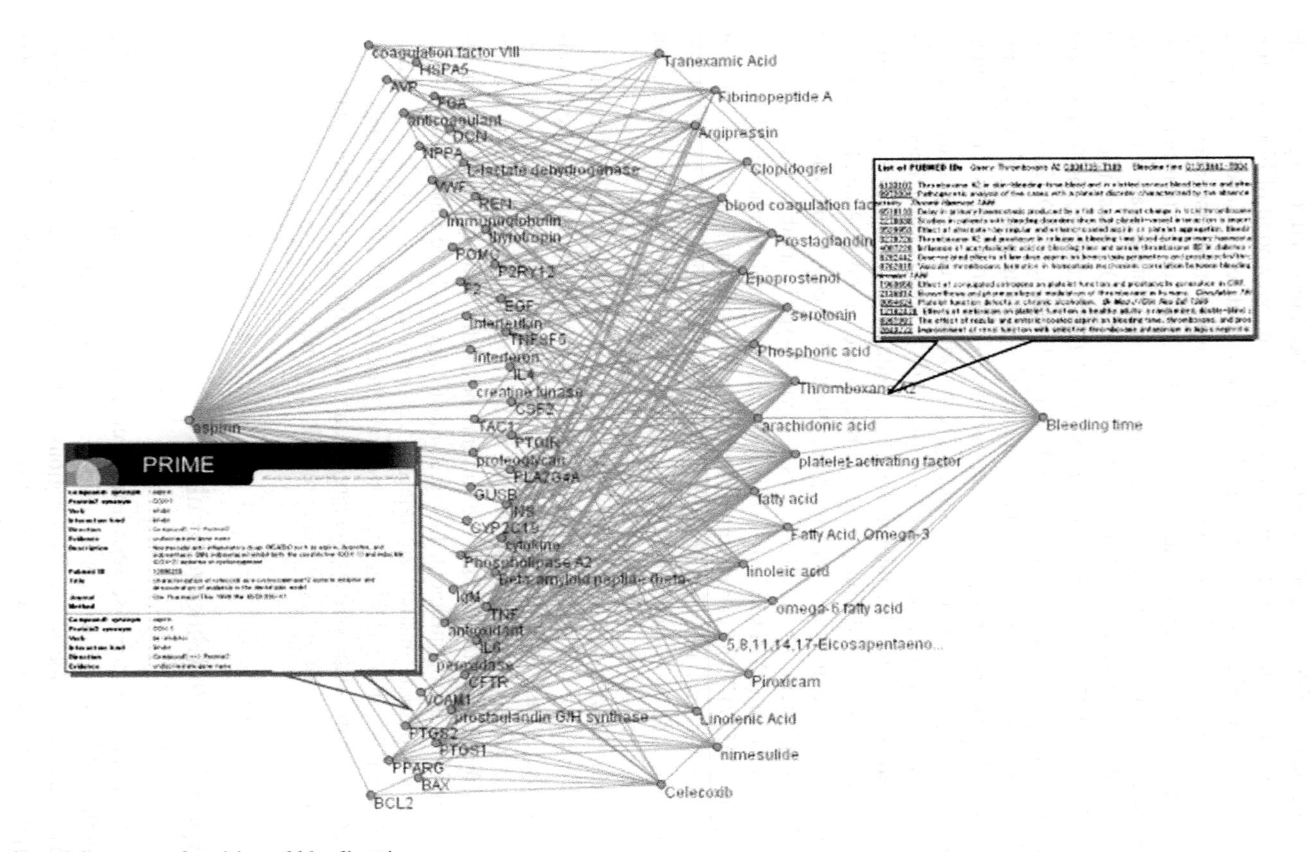

Fig. 2 Closed discovery of aspirin and bleeding time

disease-related genes difficult. For the designation of the next experiment to specify the target gene, each gene in the predicted genome region is usually investigated by reading many papers manually, and that is quite a labor-intensive task. By using BioTermNet, the target disease-related genes can be listed with the strength of relatedness and intermediate concepts.

In Table 2, the results of the application of the method to genes in the high LOD score regions of linkage analysis for myocardial infarction [30] are listed. Table 2 is calculated by open discovery with the start concept: myocardial infarction, the semantic class of end concept: gene, and one intermediate layer: semantic class of function. Documents until 2004 are used for the system validation. The used genes are located between 1p34-36 and the genome region contains about 850 genes. As shown in Table 2, top 15 predicted potential myocardial-infarction-causing genes are listed with intermediate concepts. Of these, although the direct relationship

Table 2 Summary of explicitly or potentially myocardial-infarction-causing genes

Gene name	Intermediate terms
NPPA[a,b]	Left ventricular dysfunction, brain natriuretic peptide, ejection fraction, cardiac function, neuropeptide hormone
NPPB[a,b]	Ejection fraction, cardiac death, acute coronary syndrome, neuropeptide hormone, systolic dysfunction
MTHFR[a,b]	Homocysteine, ischemic stroke, ACE, arterial thrombosis
TNFRSF1B[a,b]	Cardiac event, left ventricular dysfunction, ejection fraction
PLA2G2A[a,b]	Acute myocardial infarction of anterior wall, infarct, unstable angina
SLC2A1[a,b]	Glucose transport, cardiac muscle, arteriosclerotic coronary disease, ischemia
SLC9A1[a]	Acute myocardial infarction of anterior wall, Na + /H+ antiporter, acute myocardial infarction pathway, postinfarction
END2[a,b]	Acute myocardial infarction of anterior wall, acute myocardial infarction pathway, endothelin, angina pectoris
UTS2[a,b]	Acute myocardial infarction of anterior wall, left ventricular ejection fraction, entire coronary artery, vasoconstrictor
ECE1[a]	Acute myocardial infarction of anterior wall, endothelin-converting enzyme, acute myocardial infarction pathway
PTAFR[a,b] (Negative)	Acute myocardial infarction pathway, platelet-activating factor, physiological reperfusion, coronary artery structure
LCK[b]	Cardiac muscle, ischemic heart diseases and syndromes, 3–16 heart failure and other functional disorders, restenosis, ischemic preconditioning
ELA3A	Pancreatic function, ischemia, myoglobin, tissue plasminogen activator, steonosis, diabetes mellitus
H6PD	Glucose-6-phosphate, L-lactate dehydrogenase, NADP, glucose oxidase, anticoagulant, diabetes mellitus
GJA4[a,b]	Acute myocardial infarction of anterior wall, gap junction, arteriosclerotic, coronary, myocardium structure, left ventricular structure

[a]Indicates genes whose explicit relationships with myocardial infarction are already known
[b]Indicates genes whose functional SNPs and/or haplotypes are reported

between myocardial infarction disease and three genes (LCK, ELA3A, H6PD) were not confirmed experimentally in 2004, our system indicates their potential relationships with one intermediate functional concept layer. By clicking the edge information of graphic network, evidence to support the relationships are obtained such as "LCK is related to the ischemic preconditioning (PMID 10488057) and ischemic preconditioning is known to effect on myocardial infarct size (PMID 10588211). On the other hand, the LCK is essential in coxsackievirus B3-mediated heart disease (PMID 10742150)."

When the several physical interactions to the target genes are known, the closed discovery with the first layer: gene/protein and the second layer: function, the start concept: the targeted gene, and the final concept: myocardial infarction is also useful. Since physical interaction partners have similar biological function, the relationship to the target disease becomes clear using them as the first layer as discussed in [6]. By the addition of haplotypes or SNPs to the queries of closed discovery, genes which are related to myocardial infarction and contain already known functional SNPs or haplotype, are selected (the mark^{+} is added in Table 2).

Although BioTermNet cannot treat the function of unknown genes, at least BioTermNet is useful for planning the next steps of experiments.

6 Application to Microarray Analysis

Due to the accumulation of manually curated signaling and metabolic-pathway data, the related pathways of genes whose expression has changed are widely anticipated in microarray analysis. In BioTermNet, common concepts of predefined pathways and other semantic classes such as diseases and drug side effects with respect to expression-changed genes can be extracted. For example, when the lists of genes with expression changes induced by a drug are given, diseases to which the drug is applicable are predicted. Table 3 is the list of the diseases to which doxazosin is predicted to be applicable. Doxazosin is widely used for an enlarged prostate. Here, input is the 119 genes with significant changes in expression in response to doxazosin in a prostate cancer cell line [31] and the semantic class of second layer is a gene, and that of the third layer is a disease. The common concepts are listed by voting method, that is, the diseases having many connections (greater than a threshold) with input genes by mediating intermediate concepts are highly ranked. Since in the simple voting method, diseases with multiple disease-causing genes such as cancer are ranked too highly, the super geometric distribution based on the number of disease-related genes among input genes, the number of input genes, the number of known disease-related genes (genes that co-occur with the target disease), and the number of all genes is a more appropriate measure. The highly ranked concepts are not largely changed in this example, so the result of simple voting method is shown in Table 3. As shown in Table 3, cancer related diseases including carcinoma of prostate are highly ranked, this is because that input genes include cell growth,

Table 3 Predicted doxazosin-applicable disease list

Disease name	Indirectly related gene number
Glioblastoma	100
Neuroblastoma[a]	100
B-cell lymphoma	99
Myeloid leukemia	98
Astrocytoma	96
Fanconi's familial refractory anaemia	96
Sarcoma of prostate gland[a]	95
Adenocarcinoma of large intestine	94
Malignant tumor of ovary[a]	93
Urinary bladder carcinoma[a]	92
Autoimmune disease	93
Carcinoma of prostate[a]	91
Adenoma[a]	91
Cervical carcinoma	91
Neoplasm of thyroid gland	89
Uterine endometrial cancer	89
Inflammatory bowel disease	85
Hypertensive pulmonary vascular disease	85
Endothelial dysfunction[a]	83
Hypertension[a]	79

[a]Indicates genes on which the effects of doxazosin are already reported
Redundancy of disease names are removed manually

cell invasion, cell division, cell migration, and apoptosis related genes. The relationship between target disease and functions of input genes becomes clear when the closed discovery between the target disease and input genes is performed with semantic type of the first layer: gene and that of the second layer: function, and end concept is doxazosin as shown in Fig. 3. In this closed discovery, intermediate function concepts without direct relationship to doxazosin might be included. When the final concepts are set to be glioblastoma and doxazosin, the functions without direct relationships to doxazosin are removed. By reading the edge linked information between the function and doxazosin, the effects of doxazosin on glioblastoma are expected to be caused by the inhibitory effects on tumor cell growth (PMID: 10969806), and cell invasion (PMID: 12576871) and inducible effects on tumor cell apoptosis (PMID: 12515754).

Although benign prostatic hypertrophy (BPH), the major target of doxazosin, is not ranked within top 20 in Table 3, BPH is ranked at second (prostate cancer is ranked at first) when using the genes with significant changes in expression in response to doxazosin in a prostatic stromal cells [32].

Other than applicable diseases of a drug, prospective side effects and drug toxicity are also predictable in similar schemes.

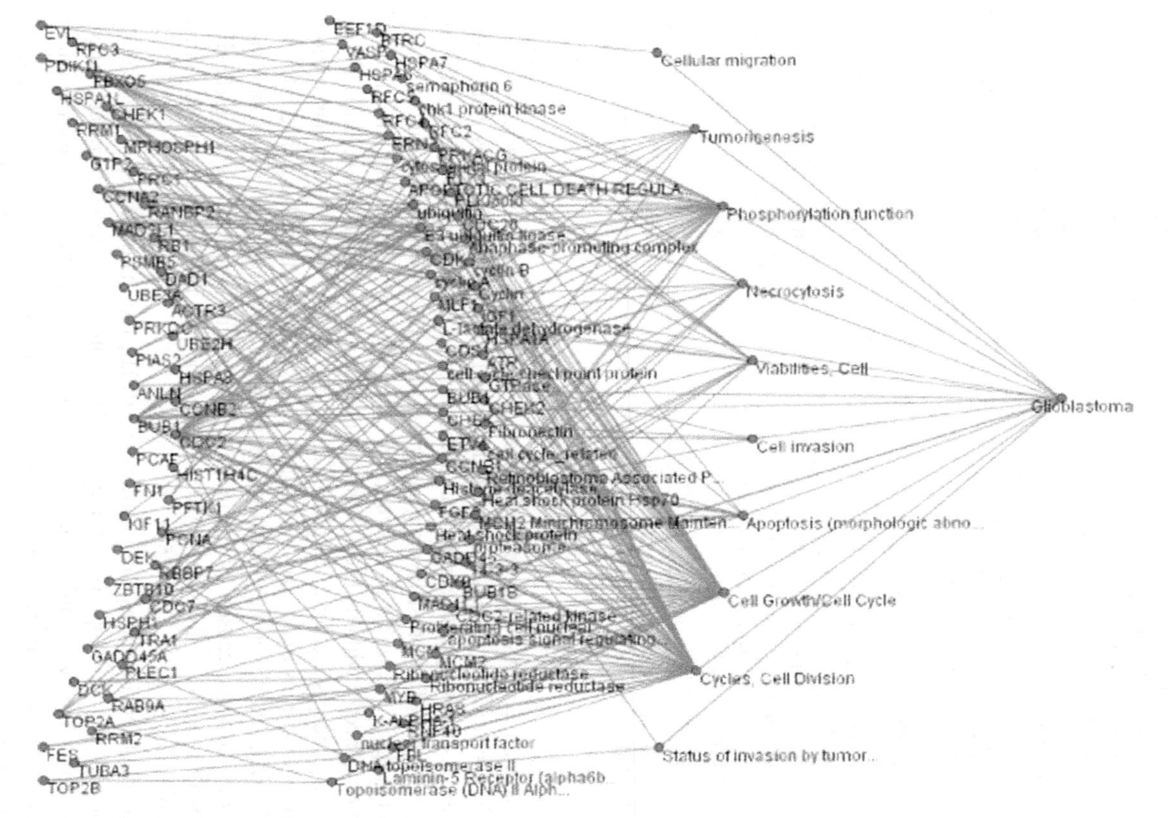

Fig. 3 Closed discovery of glioblastoma and genes with significant changes in expression in response to doxazosin in a prostate cancer cell line

7 Future Prospects

We introduced a method for knowledge discovery and its application to new drug development. In addition to the introduced subjects in this section, there is a possibility of the application of the method for the exploitation of additional indications of the approved drug, interpretation of phenotype changes caused by gene mutations and RNAi. This method is applicable to most of the subjects which uses accumulated knowledge in documents. Knowledge discovery using a crossing database is also easily performed in this scheme. Actually, knowledge discovery based on the cross-data search among the Japanese drug package leaflet data, MedLinePlus, and MEDLINE data is possible in BioTermNet, although that is not mentioned in this book. Furthermore, by combining MEDLINE data with data other than document data such as SNP frequencies and sequence information, the application of the knowledge discovery system will be expanded. The usage of full text information will enhance the potential application of knowledge discovery.

However, for the practical use of the knowledge discovery system by various researchers, the problem that an ideal system is dependent on a user's background knowledge must be addressed. The system must be evaluated by pertinence and usefulness for each user instead of system relevance measured by recall and precision. The evaluation of pertinence and usefulness is difficult in fields such as information retrieval and is even more challenging in the knowledge discovery field. The introduction of an artifice into the knowledge discovery system to accommodate the background knowledge of a user will be crucial for making the knowledge discovery system familiar to users. When it is achieved, a new paradigm of knowledge discovery will be brought out.

Acknowledgements We greatly appreciate Professor Toshihisa Takagi for his support and valuable comments. We would like to thank Dr. Tomoya Tsukahara at Hitachi East Japan Solutions, Ltd. for helping us with programming the Java viewer for BioTermNet.

References

1. Swanson DR (1986) Fish oil, Raynaud's syndrome, and undiscovered public knowledge. Perspect Biol Med 30(1): 7–18
2. Weeber M, Vos R, Klein H, De Jong-Van Den Berg LT, Aronson AR, Molema G (2003) Generating hypotheses by discovering implicit associations in the literature: a case report of a search for new potential therapeutic uses for thalidomide. J Am Med Inform Assoc 10: 252–259
3. Wren JD (2004) Extending the mutual information measure to rank inferred literature relationships. BMC Bioinformatics 5: 145
4. Van der Eijk CC, van Mulligen EM, Kors JA, Barend M, van den Berg J (2004) Constructing an associative concept space for literature-based discovery. J Am Soc Inf Sci Technol 55(5): 436–444
5. Srinivasan P, Libbus B (2004) Mining MEDLINE for implicit links between dietary substances and diseases. Bioinformatics 20(S1): 290–296
6. Koike A, Takagi T (2007) Knowledge discovery based on an implicit and explicit conceptual network. J Am Soc Inf Sci Technol 58(1): 51–65

7. Koike A, Takagi T (2004) Gene/protein/family name recognition in biomedical literature. Proceedings of HLT/NAACL BioLINK Workshop, Boston, MA, pp. 9–16
8. Koike A, Kobayashi Y, Takagi T (2003) Kinase pathway database: an integrated protein-kinase and NLP-based protein-interaction resource. Genome Res 13: 1231–1243
9. Koike A, Takagi T (2005) PRIME: Automatically extracted PRotein Interactions and Molecular NLP-based information databasE. In Silico Biol 5(1): 9–20
10. Koike A, Niwa Y, Takagi T (2005) Automatic extraction of gene/protein biological functions from biomedical text. Bioinformatics 21: 1227–1236
11. Singhal A, Buckley C, Mitra M (1996) Pivoted document length normalization. Proceedings of ACM SIGIR'96, Zurich, Switzerland, pp. 21–29
12. Fujita S (2004) Revisiting again document length hypotheses. TREC-2004 Genomics Track experiments at Patolis. Proceedings of TREC Genome Track
13. Hisamitsu T, Niwa Y (2004) A measure of representativeness based on the number of co-occurring salient words. Proceedings of the 19th International Conference on Computational Linguistics Workshop, pp. 9–16
14. Deerwester S, Dumais ST, Furnas GW, Landauer TK, Harshman R (1990) Indexing by latent semantic analysis. J Am Soc Inf Sci 41 (6): 391–407
15. Berry MW. Browne M (1999) Understanding search engines: mathematical modeling and text retrieval. SIAM, Philadelphia, PA
16. http://www.netlib.org/svdpack/
17. http://www.hprd.org/
18. Koike A (2006) Knowledge discovery and hypothesis generation from biomedical texts. Workshop on Information-Based Induction Sciences, pp. 52–58
19. http://www.ncbi.nlm.nih.gov/entrez/query.fcgi?db=OMIM
20. Ashburner M, Ball CA, Blake JA, Botstein D, Butler H, Cherry JM, Davis AP, Dolinski K, Dwight SS, Eppig JT, Harris MA, Hill DP, Issel-Tarver L, Kasarskis A, Lewis S, Matese JC, Richardson JE, Ringwald M, Rubin GM, Sherlock G (2000) Gene ontology: tool for the unification of biology. The Gene Ontology consortium. Nat Genet 25(1): 25–29
21. Homayouni R, Heinrich K, Wei L, Berry MW (2005) Gene clustering by latent semantic indexing of MEDLINE abstracts. Bioinformatics 21(1): 104–115
22. Swanson DR (1988) Migraine and magnesium: eleven neglected connections. Perspect Biol Med 31(4): 552–557
23. Swanson DR (1990) Somatomedin C and arginine: implicit connections between mutually isolated literatures. Perspect Biol Med 33: 157–186
24. Smallheiser NR, Swanson, DR (1996) Indomethacin and Alzheimer's disease. Neurology 46(2): 583
25. Smallheiser NR, Swanson DR (1998) Calcium-independent phospholipase A2 and schizophrenia. Arch Gen Phychiatry 55(8): 752–753
26. Hankey GJ, Eikelboom JW (2004) Aspirin resistance. BMJ 328(7438): 477–479
27. Maree AO, Curtin RJ, Chubb A, Dolan C, Cox D, O'Brien J, Crean P, Shields DC, Fitzgerald DJ (2005) Cyclooxygenase-1 haplotype modulates platelet response to aspirin. J Thromb Haemost 3(10): 2340–2345
28. Pamukcu B, Oflaz H, Nisanci Y (2005) The role of platelet glycoprotein IIIa polymorphism in the high prevalence of in vitro aspirin resistance in patients with intracoronary stent restenosis. Am Heart J 149(4): 675–680
29. http://www.hapmap.org
30. Wang Q, Rao S, Shen G-Q, Li L, Moliterno DJ, Newby LK, Rogers WJ, Cannata R, Zirzow E, Elston RC, Topol EJ (2004) Premature myocardial infarction novel susceptibility locus on chromosome 1P34-36 identified by genomewide linkage analysis. Am J Hum Genet 74: 262–271
31. Arencibia JM, Del Rio M, Bonnin A, Lopes R, Lemoine NR, Lopez-Barahona M (2005) Doxazosin induces apoptosis in LNCaP prostate cancer cell line through DNA binding and DNA-dependent protein kinase down-regulation. Int J Oncol 27: 1617–1623
32. Zhao H, Lai F, Nonn L, Brooks JD, Peehl DM (2005) Molecular targets of doxazosin in human prostatic stromal cells. Prostate 62(4): 400–410

Index

A-Terms, 102
ABC model, 5, 6
Additional sources within the LBD process, 79
Algebra of sets, 6
Alu elements, 14, 16, 17
Ambiguity challenge, 96
Analysis of gene-disease LBD approaches, 85
Analysis of the General Purpose LBD Systems, 90
Application of Swanson's Open discovery approach, 92
Apriori algorithm, 87, 94
ARROWSMITH, 87
Assimilation of knowledge, 7
Autoimmune disease database, 83
Automated indexing, 158
Automatic relevance predictions, 19

B-terms, 9, 102
Background knowledge, 5
Balliol College, 7
Baseline set, 107
"Benchmark" LBD problems, 89
Betweenness centrality, 84
Bibliographic content, 155
Bibliographic databases, 6–10
Bio-SbKDS, 106
Bioinformatics text mining, 79
Biology, 7, 8
Biomedical scientist, 8
BIOSIS, 8
BioTermNet, 173
BIOTLA, 82
Bipartite topic sets, 89
Bird's eye view of LBD research, 76
BITOLA, 107
Blair, D, 10
Blogs, 95
Boolean operators, 165
Boolean searching, 164
Boveri–Sutton hypothesis, 3
Breakthrough, 7
Building block, 7

C-Term, 102
Candidate terms, 8
Cardiac hypertrophy, 118, 125–127
Cartoon, 4
Case study, 14, 17
Cell division, 3
Centers, 93
Chilibot, 85
Chromosome behavior, 3
Citation indexes, 7, 9
Citation linkages, 61
Citation pattern, 7
Clinic, 8
Closed discovery, 78
Closed discovery approach, 106
Closed discovery system, 177
Closed end search, 8
Co-occurrence, 84, 85, 90
Co-occurrence based analysis, 89
Co-occurrence based connections, 78
Co-occurrence based network, 95
Co-occurrence frequency, 89
Co-occurrence network, 84, 95
Combinatorial growth, 8
Communities of genes, 84
Complementarity; exemplar, 3
Complementarity; suggestive, 5
Complementary but disjoint (CBD), 9, 10
Complementary structures, 9
Conference web sites, 95

Confidence and support scores, 82
Connection explosion, 8
Controlled vocabularies, 158
Convergence, divergence, 94
Core, 93
Cosine coefficient, 179
Cosine normalization, 89
Cosine similarities, 95
Crohn's disease, 63, 64
Cross citation, 7
Cthulhu, 10
Curcumin, 63–65
Cut-off date, 107
Cytogenetics, 3, 4
Cytology, 3, 4, 7

DAD, 105
Dark age, 10
Data (e.g., systematic reviews and/or meta-analyses), 154
Data Integration, 52
Date 1866, 3, 7
Date 1900, 3, 4, 7
Defining LBD success, 17, 20
Derwent innovations index, 64
Dice coefficient, 179
Digital libraries, 153
Discovering connections between people, 91
Discovering novel links from the Web, 91
Disjoint, 9, 10
Dispersed knowledge, 8
Dissociated knowledge, 10
Division of labor, 7
Domain-independent meta-analysis of LBD research, 76
Dominant trait, 4
Dublin Core Metadata Initiative (DCMI), 161

Eigenvalue decomposition, 118, 119, 123
Electronic publication, 157
End-users, 18
Entrez Gene, 77
Evaluation, 101
Evaluation of IR systems, 166
Evidence combination models, 86
eVOC Anatomical System ontology, 81
eVOC system, 81
Evolutionary improvement, 8
Exact-match searching, 164
Exploratory searching, 8
Extract semantic links between entities, 92

F-Measure, 103
Factor analysis, 118, 121–128
Factor screening, 119–121, 127
False negatives, 104
False positives, 103
Fans, 93
Fertilization, 3, 4
Field testing, 13, 17–19
Finding connections involving companies and industries, 91
Form and/or explore hypotheses, 76
Fragmentation of science, 7
Frequent itemsets, 94

G2D, 79
Gardner-Medwin, 8, 10
Garfield, Eugene, 7
GENA, 174
Gene Ontology, 77
Gene–disease links, 79
Gene–disease problem, 82
General framework, 76
General purpose biomedical text mining systems, 87
Generate novel ideas, 77
Geographical co-location, 79
Gold standard, 18, 19, 102
Graph partitioning algorithm, 84

Hereditary transmission, 4, 7
Heredity, 4
HG, 179
High-recall, 9
Historical discovery, 50
Hubs, 93
Human capacity, 7
Humanities domain, 92
Humanities index, 93
Hybridity, 3, 4, 7
Hypergeometric distribution, 104
HyperGsum (HG), 179
Hyperlinks, 91
Hypothesis, 3, 8, 10
Hypothesis follow-up, 20
Hypothesis generation, 129, 173

Idealization, 9
Identifying relationships between companies, 94
Implicit connections, 8, 78
Implicit discovery, 6
Implicit knowledge, 7
Implicit relationship, 41, 42, 46–48, 50, 117
indexing, 155
Inference methods, 78
Inference validity, 50

Information explosion, 8
Information extraction, 92
Information extraction techniques, 94
Information retrieval, 9, 102, 153
Inlinks, 93
Innovation, 57, 58, 60, 61, 63–66, 69
Insular scientific communities, 7
Intermediary literature, 116, 117, 119, 120, 122–125, 127–129
Inverse document frequency (IDF), 163
IR, 102
IRIDESCENT, 45, 46, 87

Journal article, 7

Knowledge discovery, 40, 46, 50–52, 133, 135, 137, 139, 141, 143, 145, 147, 149, 151, 173
Knowledge discovery from text (KDT), 75
Knowledge-based information, 154

Latent semantic indexing (LSI), 180
LBD, 101
Lexical-statistical retrieval, 165
Linkage analysis, 185
Linking terms, 102
Links, 78
Literature based discovery, 75, 77
Literature of science, 8
Literature structure, 3, 4, 6
Literature-based discovery, 101, 133, 134, 139, 150, 151
Literature-Based Discovery Systems, 101
Literature-based resurrection, 9
LitLinker, 87, 107
LitMiner, 83
Logical statement, 5
Lovecraft H.P., 10
LSI, 180

Manjal, 89, 106
Manual indexing, 159
Matrix decomposition, 117, 118, 128, 129
Mean average precision, 167
MedGene, 83
Mediator sequences, 94
Mediator sets, 94
Medical Subject Headings (MeSH), 159
Medicine, 7, 8
MEDLINE, 62–64, 66, 77, 155
Medline, 8, 59, 61, 65
MEDLINE analysis, 40, 41, 43, 44, 52
Medline descriptors, 114
Meiosis, 4
Mendel G, 3, 4, 7, 9
MeSH concepts, 82
MeSH profiles, 89
MeSH terms, 59, 62–65
MI, 179
MI* freq, 179
MicroRNA, 14–18
Migraine, 3, 6, 7, 10, 116–119, 123–125
Migraine source literature, 124
MIM, 108
Multiple sclerosis, 118, 127, 128
Mutual information (MI), 88, 179

Natural language context/situation/usage, 5
Natural language processing, 46, 134, 136, 138, 150
Natural language query, 165
Neglect, 9
Network analysis, 94
Network of associated objects, 90
NLP, 51
NLP methods, 78
Non-traditional “documents”, 92
Novel links between web pages, 94
Novel relationships between person entities, 92
Novelty, 6
Null intersection, 9

Objects, 77
OMIM, 77, 174
Online communities, 93
Open discovery, 78
Open discovery approach, 105
Open discovery model, 41
Open Discovery System, 178
Open end search, 8
Open questions and directions for research, 97
Outlinks, 93

PageRank, 163
Partial-match searching, 164
Pattern recognition, 78
Pea hybridization, 3
Peas, 4
Perl, 118
Pharmaco genomics, 184
Precision, 102, 167
Precision-recall graphs, 103
Profile similarity, 89
Protein–protein interactions, 85
Pubmed, 155

Radical discovery, 58–60
Raynaud's disease, 59, 61, 62, 69, 116–118, 121–123, 135, 140
Raynaud's Disease, Migraine, 119
Real world, 8
Recall, 102, 167
Recall-precision table, 167
Receiver Operating Characteristics (ROC), 103
Recessive trait, 4
Recombine, 4
Recorded knowledge, 7
Relationship Discovery, 41, 46–48, 50, 52
Relationship evaluation, 42
Relationship Prioritization, 48
Relationships, 46
Relationships between people, 92
Resource Description Framework (RDF), 162
Restructuring web sites, 94
Retrieval, 155

Science Citation Index, 61, 64, 158
Scientific journal, 7
Scientific knowledge; edifice, 7
Scientific problem, 3
Search command language, 6, 8
Search logs, 95
Search strategies, 10
Semantic groups, 108
"Semantic" links, 90
Semantic relation extraction, 145, 149, 150
Semantic space, 117, 128
Semantic Web, 162
Semantically motivated relationships, 78
Sensitivity, 103
Sets of articles, 9
Shared research techniques, 4
Similarity-based assessments, 78
Simple framework for analyzing LBD methods, 77
Simpson coefficient, 179
Single paths versus multiple paths, 90
Singular value decomposition (SVD), 117, 179
Software package R, 118
Source literature, 116, 121, 122, 124, 126, 129
Specialization, 8
Specialization in science, 7
Specificity, 103
Spreading cortical depression, 6
Starting term, 102
State of mind, 6
Stemming, 163
Stop list, 163
Summing up, 10
Sutton, Walter, 3, 4, 7
Sutton-Boveri hypothesis, 3
Swanson linking, 116
Syllogism, 5
Symmetric versus asymmetric inferencing strategies, 91
Symmetric versus asymmetric methods, 86
System comparisons, 103

Target literature, 116, 129
Target terms, 102
Temporal analysis, 93
Term frequency (TF), 163
Test collections, 167
Test set, 107
Text mining, 75, 115–117, 129
Text REtrieval Conference, 167
TF*IDF weights, 89
TF-IDF, 179
Thesaurus, 159
Topic profiles, 89
Transcription factor associations, 85
Transitive, 115
Transitive analysis, 93
Transitive relationship, 78
Transitive structure, 92
Transitive text mining, 116
Transitivity, 94
Trial-and-error, 8
True negatives, 103
True positives, 103
Truncation, 165

UMLS, 174
Undiscovered public knowledge, 6, 10
Undocumented, 9
Unified Medical Language System (UMLS), 174
Unintended, 9
Unknown, 9
User Interface Evaluation, 111
User-access data for Web sites, 95
User-defined problem, 8, 10

Validation of Potential Discovery, 61, 62
Value of LBD predictions, 17
Vector-space model, 165

Web catalog, 156
Web of Science, 158
Web pages, 91, 95
Wild-card character, 165
WILS database, 93
Wittgenstein, 10

Z-Score, 108
Ziman J.M. Public Knowledge, 11